Process
Management

Other McGraw-Hill Quality Control Books of Interest

CARRUBBA · *Product Assurance Principles: Integrating Design and Quality Assurance*

CROSBY · *Let's Talk Quality*

CROSBY · *Quality is Free*

CROSBY · *Quality Without Tears*

FEIGENBAUM · *Total Quality Control, Third Edition, Revised, Fortieth Anniversary Edition*

GRANT & LEAVENWORTH · *Statistical Quality Control*

JURAN & GRYNA · *Juran's Quality Control Handbook, Fourth Edition*

JURAN & GRYNA · *Quality Planning and Analysis*

MENOD · *TQM in New Product Manufacturing*

MILLS · *The Quality Audit*

OTT & SCHILLING; · *Process Quality Control*

ROSS · *Taguchi Techniques for Quality Engineering*

SLATER · *Integrated Process Management: A Quality Model*

SAYLOR · *TQM Field Manual*

TAYLOR · *Optimization and Variation Reduction in Quality*

Process Management

Methods for Improving Products and Service

Eugene H. Melan
Marist College

Copublished with ASQC Quality Press

McGraw-Hill, Inc.

New York St. Louis San Francisco Auckland Bogotá
Caracas Lisbon London Madrid Mexico Milan
Montreal New Delhi Paris San Juan São Paulo
Singapore Sydney Tokyo Toronto

Library of Congress Cataloging-in-Publication Data

Melan, Eugene H.
    Process managment : methods for improving products and service
    Eugene H. Melan.
        p.   cm.
    Includes index.
    ISBN 0-07-041339-8
    1. Production management—Quality control. 2. Process control.
    I. Title.
    TS156.M433   1992
    658.5'62—dc20   92-27834
                                                       CIP

1 2 3 4 5 6 7 8 9 0 DOC/DOC 9 8 7 6 5 4 3 2

ISBN 0-07-041339-8

The sponsoring editor for this book was Gail F. Nalven, and the production supervisor was Donald F. Schmidt. This book was set in Century Schoolbook by Publication Services, Inc.

Printed and bound by R. R. Donnelley & Sons Company.

Contents

Preface   viii

Part 1   Fundamentals of Process Management   1

Chapter 1   Introduction   3

Evolution of Organization   5
The Service Industry Evolution   8
Quality and Market Share   9
Quality Improvement and TQM   10

Chapter 2   Origins and Characteristics of a Process   13

Process Elements   14
Transformation Model   16
Characteristics of a Process   20
Service Processes   23
Work Flow and Organizational Structure   24

Chapter 3   Fundamentals of Process Management: Process
Initialization   27

Process Ownership   27
Boundaries and Interfaces   31
Customer–Producer–Supplier Model   33
Managing the Interface: A Matrix Method   41

Chapter 4   Fundamentals of Process Management: Defining the
Process   45

Flowcharts and Symbols   46
Defining a Process   48
Drawing Flowcharts   54

Chapter 5   Fundamentals of Process Management: Process Control   59

Control Points   59
Reactive and Controlled Processes   60
Methods for Service Process Control   63
Measurements   65

Considerations in Selection and Implementation   70
Graphical Methods   72
Feedback and Corrective Action   77

Chapter 6   Analyzing the Process: The Classical Method   81

Process Analysis   81
Process Analysis Procedure   82
Application Example   82

Chapter 7   Analyzing the Process: Modern Methods   87

Department Work Product Analysis   87
Cross–Functional Process Analysis   98
Service Blueprint   105

Chapter 8   Assessing and Evaluating a Process   111

Criteria for Assessing a Process   112
Process Evaluation: Rating Method   115
Adaptability   119
Other Evaluation Methods   119
Appendix: Process Assessment Questionnaire   125

Chapter 9   Process Management in Practice I: Putting It All Together   127

Step 1: Establish Ownership   128
Step 2: Establish Boundaries and Interfaces   129
Step 3: Define the Process   130
Step 4: Establish Customer Requirements   131
Step 5: Establish Control Points   132
Step 6: Measure and Assess the Process   133
Step 7: Perform Feedback and Corrective Action   136

Chapter 10   Process Management in Practice II: Implementing Process Management for TQM   139

TQM and Its Features   139
Improvement Through Process Management   149
Examples of Process Management Implementation   155

Part 2   Cases in Process Management Applications   165

Chapter 11   Process Management in Staff/Service Operations   167

Case 1: An Ordering Process: A Case of Late Delivery   167
Case 2: A Management Reporting Process: A Case in Wasting Resources   177
Case 3: Analysis of a Transportation Service Process: A Case in Managing the Interface   184

Chapter 12   Process Management in Financial Operations   193

Case 1: Accounts Payable: A Case of Late Payments   196
Case 2: Accounts Receivable: A Case of Timely and Accurate Billing   201

Chapter 13   Process Management in a Laboratory   209

Development Phases   210
Case 1: The Development Process: Cycle Time Analysis   213
Case 2: A Service Process: Methods for Improving Service Quality   215
Case 3: Engineering Services: A Process Team Approach to Improvement   221

Part 3   Processes: Present and Future   227

Chapter 14   Designing a Process   229

Requirements for Process Design   230
The Concept of Work Flow   234
Constructing a Process   237

Chapter 15   Future Trends   247

The Quality Revolution   247
Organizational Trends   249
A New View   255

Index   257

Preface

The concept of total quality management (TQM) has become widely publicized in recent years, particularly through the efforts of industry and professional organizations such as the American Society for Quality Control. The federal government through the Commerce Department has taken a lead in fostering quality improvement in industry through the Malcolm Baldrige Award for quality, which has become the American equivalent of Japan's Deming prize. As a result of the upsurge in quality, there is an increasing interest in techniques and methods for improving quality.

Bookshelves of quality improvement practitioners are replete with books on statistical methods, the Deming philosophy, Japanese quality methods and techniques, quality control, and participative management methods. But, to my knowledge, there is little available that provides an insight into the theory and application of process quality, the subject of this book.

This book, which is based on over eight years of research, teaching, and practice, is written from the point of view of the theory and implementation of process quality improvement through the concept of process management. It is written with both the uninitiated as well as the experienced TQM practitioner in mind. It is written to provide an understanding of the concepts and how they can be applied.

The concept involves applying certain principles in analyzing and improving the work flow of an operation. Process management has evolved from the realization that techniques of managing work flow in manufacturing can also be applied to nonmanufacturing operations of a business. Process management has proven to be an effective way to examine a business process; many operational improvements have been achieved with this approach with relatively little investment in resources. The intent of this book is to provide the reader with a set of tools and working concepts to achieve a greater insight and understanding of the business processes and how to improve them.

Viewing a business operation as a process has become an important concept in recent years. Applying this concept has resulted in significant improvements in cost, productivity, and quality in many business areas. Areas such as finance, purchasing, warehousing, and marketing have employed it with excellent results.

The book is divided into two parts. In Part I, the origins and growth of industrial organizations and the evolution of administration and bureaucracy are presented. The notions of process and work–flow are discussed with their general characteristics and features. Because every process is contained within an organization or a group of organizations, one of its key characteristics, suboptimization, is highlighted. A means for addressing suboptimization, namely the customer-producer-supplier relationship model is then described.

The paradigm of process management is best seen by examining the general features of a well-managed manufacturing process, Chapter 2, and contrasting them with that of typical administrative and service processes. These features form a basis for the following three chapters, which describe the three phases of process management: initialization, definition, and control.

In the initialization phase, Chapter 3, the precept of process ownership, which is vital to managing and improving processes, is discussed. It is at this point that we distinguish the human or social dimension from the technical dimension of a process. Fundamentally, all processes operate within a framework of a sociotechnical system. I return to this notion in Part II during the discussion of the elements of designing a process. This chapter also examines the concepts of boundaries and interfaces, which serve to delimit the process and identify critical points of the work flow as it passes from a producer to a receiver.

In the definition phase, Chapter 4, the symbolic method of portraying a flow of work activities, namely flowcharting is reviewed. In Chapter 5 the final phase, control, is described. Control consists of three steps: establishing points of control, performing measurements, and regulating. Control points provide a means for assessing activities that have an influence on the outcome of a work product—verification, for example. Measurements provide a basis for feedback regulation and improvement. In turn, regulation is intended to manage process variability and also serves as a basis for control.

In Chapters 6 and 7 various methods for defining and analyzing a process are described. The first of these two chapters contains the traditional, or classical, method of process analysis. The second examines more recently developed methods: department or work–group analysis, cross–functional analysis embodying the highly formalized process analysis technique used for complex processes, and a method developed specifically for service processes, the service blueprint.

With the theory of process management in hand, we now turn to assessing and evaluating a process. The intent of Chapter 8 is to provide the reader with the means of evaluating an existing process, as often occurs in TQM work. Techniques such as the rating method, performance evaluation, benchmarking, and the quality profile are described.

In Chapter 9 theory is put into practice through examination of a simple process found in many administrative operations, namely, document distribution. Here we apply many of the concepts developed in the earlier part of the book. Part 1 is concluded by showing how process management is implemented in a TQM framework and how the methodology has been applied at NCR and AT&T.

In Part 2, various applications of process management are shown. We begin with three cases illustrating administrative and service processes in Chapter 11, showing first how a complex, cross–functional process such as ordering can be dissected and improved, followed by a case in budget management and a case in service—delivery and pickup of packages. These cases, though seemingly divergent, illustrate common problems among business processes.

In Chapter 12 two processes common to any financial operation, accounts payable and accounts receivable are analyzed. Chapter 13 presents three cases illustrative of product development: a design process which represents a microcosm of a larger design process and two engineering service processes examined from opposite points of view—the quality of delivering the service and the quality of performing the service.

Lastly, in Part 3 process design is addressed (Chapter 14). Although process management is generally applied to existing processes, it is important to consider how processes can be designed from the beginning and what must be taken into account in putting a process together. In the concluding chapter, I point out to the reader certain business developments of this decade that affect the way work is addressed. These developments point to an increasing recognition of the need to adopt the systems–process view of our world. The ending remark, a quotation by Peter Drucker, summarizes the importance of this view: "The process produces results."

I am deeply indebted to a number of people who have made this book possible. First, I would like to thank Edward F. Sullivan, former editor of *Quality Progress*, for his encouragement in writing a book on this topic, his invaluable editorial guidance on the early drafts, and most importantly for initiating the ASQC Business Improvement Symposia which was instrumental in the spread of the concept throughout the United States. I also thank Edward P. Leonard, Science Research Associates, and Frank Topper, University of Pennsylvania, for their reviews and invaluable suggestions from the "customer" viewpoint; Dr. John C. Kelly, Chairman of the Management Studies division of Marist College, and Dr. Dennis Murray, President of Marist College, champions of TQM, for their suggestions, ongoing encouragement, interest, and most importantly, a supportive environment conducive to research and intellectual growth.

Two individuals deserve special note. David P. Kennedy, Vice President for Quality, Grand Metropolitan, is the person who, while a member of the IBM corporate quality staff, brought the systems view to the attention of IBM executive management and sowed the seeds for what ultimately developed into the concept of process management and IBM's Quality Focus on the Business Process. Finally, this book would never have been accomplished without the patience, perseverance, and unwavering dedication of an unsung hero, Beverly C. DeMott, in the typing and editing of this work from its inception over six years ago.

Eugene H. Melan
Poughkeepsie, New York

Fundamentals of Process Management

Introduction

The concept of total quality management (TQM) has gained widespread popularity in recent years. From its original application to manufacturing firms, it gradually spread to the service industry, and recently there is strong evidence of TQM activity in government and education as well.

Fundamental to TQM is understanding and improving the operations of an organization—its processes. Examining the business operations of an organization and showing how they can be improved are the central themes of this book. The results of improving the way we do business are indisputable: higher quality of the product or service, increased productivity, improved operational efficiency, and lower costs. A typical example of this improvement is what was achieved by the Intel Corporation in California in its accounts payable operation. The processing of expense vouchers, vendor payments, freight invoices, and expense accounts was examined. As a result, various activities were simplified and improved. These improvement efforts enabled Intel to reduce its accounts payable staff from 71 to 51 in less than a year while improving its voucher processing time by 30 percent.[1] Opportunities such as this for reducing administrative costs, improving productivity, and eliminating waste exist in many companies today, but they are often unrecognized or ignored by management.

The magnitude of waste occurring in business operations is mind boggling. It has been estimated that bureaucracy accounts for over $800 billion in waste a year—six times the waste incurred in the federal government![2] Much of this occurs in the administration of business and much of it is avoidable. In recent years magazine and newspaper articles have cited many examples of swollen staff, excessive layers of high-salaried management contributing little or no added value to the firm, ponderous and time-consuming decision-making procedures, and expensive perquisites of rank approaching that of royalty.

Waste is preventable. Unfortunately, however, it is often blindly accepted. Witness the following examples:

- The production line of a small, high-tech company in the northeastern United States produces magnetic components used in computer storage devices. Its overall yield has hovered around 50 percent for nearly two years in spite of sporadic attempts at improvement. Management recognized that over half of the yield loss resulted from mishandling these components while they are being made. The cost of scrap and rework was calculated to be $600,000 a month. Various excuses were offered for not being able to achieve higher yields even though management admitted that yields in excess of 70 percent were possible. The firm, facing Chapter 11, was bought out by a company based in the Far East and was restructured to one-third of its original size.

- A representative of a prominent European manufacturer of prestigious crystal products proudly claims to visitors touring its manufacturing operations that 45 percent of what it makes is destroyed. When asked why so much is destroyed, the representative explains that it is because of the delicate nature of the glassmaking operations and the manufacturer's high quality standards. This company has reported significant operating losses in the past two years.

- The management of a development organization of a large business equipment manufacturer believes that there is no practical way to reduce its product design cycle of 48 months even though competition is achieving significantly shorter development cycles. When asked why, the response was, "We do things differently." This firm has encountered a substantial loss in market share and profits due to competition.

- The administration of a large university in the northeast accepts duplication of various activities among the different schools within the university as "the price one pays for a decentralized organization structure." This institution faces appreciable reductions in state and federal funding.

These are but a few examples of the problems facing organizations today. In all of these examples a common theme emerges: a tacit acceptance of waste and inefficiency even though improvements are necessary for survival. In many cases, management does not have an insight into the operations it is accountable for. There is an unquestioning acceptance of complexity and a reluctance to understand details. Management waits for a crisis to force it to take action—oftentimes too late.

Given the fact that U.S. industry once had a worldwide reputation for efficiency and productivity, one wonders how inefficiency and waste has occurred even in the most well-managed companies. To see how this has come about, let us first examine how industrial organizations evolved to their present state.

Evolution of Organization

The origins of industrial organizations began when the concept of the division of labor was developed. Early evidence of labor specialization can be seen in the wall art and hieroglyphics of ancient Egypt. A tomb of a pharaoh built around 1800 B.C. shows a worker measuring the diagonal dimension of a stone block while another worker is chiseling it. Stoneworkers in ancient Greece had their tools sharpened by other workers whose only job was to sharpen tools, enabling them to focus exclusively on shaping the stones. Early industrial history is replete with examples of technology creating skill specialties such as glass blowing and wood carving.

By the middle of the eighteenth century Adam Smith, in *The Wealth of Nations*, espoused the advantages of specialization and provided an economic foundation for the Industrial Revolution. Smith saw that specialized labor provided increased output through improved productivity and gave a lower cost of manufacture—advantages that our early entrepreneurs understood. Pay commensurate with the type of skill became the basis for the reward system. An economic basis for specialized labor activity now existed: creating products efficiently. This in turn provided a rationale for functional organizations.

The Industrial Revolution itself created the factory system. As machine technology developed, resources had to be concentrated to produce goods efficiently on a large scale. Cottage industries became obsolete as factories, both large and small, replaced them. In the United States, the period after the Civil War represented a transition from small enterprises to the factory system. Operations became at once more specialized and more complex. With growth, organizations became further compartmentalized, resulting in structures that were both functional and hierarchical in nature. Chandler, a prominent business historian, describes pre–Civil War business organizations in this country:

> Before 1850, very few American businesses needed the services of a full-time administrator or required a clearly-defined administrative struc-

ture. Industrial enterprises were very small in comparison with those of today. The two or three men responsible for the destiny of a single enterprise handled all the basic activities—economic and administrative, operational and entrepreneurial.... Merchants, manufacturers or railroad officers spent nearly all their time carrying on functional activities—the actual buying and selling or the personal supervision of the operations of a mill or a railroad.[3]

He points out that a few of the largest businesses at the time developed administrative operations. The Second Bank of the United States, the American Fur Company, and the largest railroads of the time (Erie, New York Central, Pennsylvania, and Baltimore and Ohio) contained headquarters functions such as finance, accounting, and purchasing. Railroads became the catalyst for growth of American firms engaged in mining, manufacturing, and marketing. As these firms expanded, effort was required to coordinate, plan, and appraise the line organizations, which inevitably led to the creation of staff and administrative operations.

Chandler also notes that, starting in the 1880s, the major impetus to establishing large administrative structures was based on the creation and absorption of new functions rather than simply on an increase in output. Manufacturing enterprises created their own marketing functions and merchandising companies moved toward controlling or owning manufacturing plants—strategies of forward and backward integration. Tobacco and meat packing became prime examples of vertically integrated, multifunctional organizations.

Simultaneous with business strategies of forward and backward integration were developments in horizontal combinations and consolidation of firms. Consolidations of the 1880s and 1890s in the oil business are illustrative of this development. These combinations also provided an impetus to vertical integration in order to achieve lower product costs; little or no attention was paid to general operational efficiency. Administrative staffs became institutionalized. Management also came to believe that size was the key to market share.

As these enterprises evolved, so did organization. Specialized departments such as bookkeeping, personnel, payroll, and shipping were created to address the internal needs of the business. Over time more intricate business structures developed. These operations often contain webs of interrelated activities so complex that they stagger the imagination. A billing process, for instance, may have over fifty distinct activities from beginning to end. These complex processes are embedded within organizations that prevent them from operating efficiently because of the bureaucratic nature of the structure itself. This phenomenon has been termed the *hidden plant* and results in nonproductive activity and the waste we accept.[4]

Functions such as finance, personnel, and engineering evolved within a pyramid or command type management structure and, although their objectives were to help produce a product or service, they were not generally integrated directly into the manufacturing function. These functions operated within an organization that was not designed to maximize the production process. As a result, conflicts in functional objectives occurred that led to suboptimization* and inefficiencies—characteristics which exist in many of today's enterprises, large and small. This hidden plant represents a significant internal cost of doing business in many organizations—both profit and nonprofit. The results are waste of resources, dissatisfaction with the end product or service, and frustration on the part of everyone— customers, managers, and employees. Even worse, a firm's competitive capability is reduced and it is more likely to suffer from reduced revenues as business downturns occur.

The MIT Commission on Industrial Productivity has pointed out that lack of cooperation of individuals within many U.S. firms has been a significant detriment to productivity improvement of those organizations.[5] The Commission points out that many firms have organizational barriers or "walls" that separate product development from manufacturing and marketing from both development and manufacturing. Because of these organizational barriers, employees are prevented from working as interfunctional teams. This lack of cooperation has come about from compartmentalization of individual and group functions and resultant narrow self-interest.

Another aspect of the hidden plant is what Rosabeth Moss Kanter calls *segmentalism*—the compartmentalizing of events and actions in an organization while keeping each area isolated from others.[6] She points out that organizations with segmentation embedded in their cultures are likely to have segmented departments and functions isolated from one another, with minimal interaction between them and little incentive to cooperate with one another.

---

*The term suboptimization refers to management decisions and behavior that may optimize the use of resources within the function itself but may be detrimental to the operation of the entire firm. For example, a manager within a function may choose to eliminate overtime for budgetary reasons. This may cause delays in shipping a product to a customer, which will result in cancellation of an order—a sales loss. Suboptimization often occurs because work performance and behavior are rewarded based on what the individual or group does for the organization itself rather than what is done for the total productive system.

Suboptimization, lack of cooperation, and segmentation, then, are features of the bureaucratic nature of organizations and represent forces working against productivity, operational efficiency, and effectiveness.

As business enjoyed the apparent economic benefits of size and overlooked or endured bureaucratic inefficiency, its complexion began to change. Increased competition of foreign products brought a decline in production of many domestic consumer and industrial goods while service and knowledge-based industries began to grow dramatically. The United States began to evolve toward a service-oriented, information-based society. Service became its fastest-growing industry in the past decade.

The Service Industry Evolution

Government statistics indicate that 68 percent of the nation's GNP and 71 percent of the work force are involved in the service sector. It has been estimated that 90 percent of the working population will be in service industries by the end of the century. The $2.3 trillion private services sector is largely composed of transportation, utilities, communications, retail and wholesale trade, finance, insurance, and real estate, with the latter two categories comprising over half the total. If government service is included, an overwhelming majority of the U.S. work force is, and will be, connected with the service sector.

Analysis of this sector shows, surprisingly, that services are substituting directly for manufactured products across a wide spectrum.[7] Many services, such as information systems and financial services, are replacing traditional products. The economic value of the service sector is at least as high as that of manufacturing; in some cases it is even higher.

Also of significance is that support services have become a critical cost dimension of a country's competitive posture in manufacturing. They can substantially lower the real cost of producing goods in such areas as transportation, communications, finance, insurance, and health care. Therefore, eliminating the hidden plant and providing efficient support services in these areas can be instrumental in achieving competitive product costs. For many manufacturing companies, then, service improvement is becoming a means of obtaining competitive advantage. Not only do those services afford a competitive edge, but they also provide profitability, because the more service-sensitive the market the higher the return on both sales and investment.

Quality and Market Share

Business executives have also begun to realize that quality has become a crucial factor in their businesses. A Gallup survey of executives of *Fortune* 500 and other manufacturing/service companies, sponsored by the American Society for Quality Control (ASQC), indicates that 41 percent of them ranked product and service quality as the most critical issue facing American business in the next three years.[8]

Some businesses have found that improving product quality is a powerful way to improve market share. For consumer and industrial goods, firms that increased the relative quality of their products achieved greater gains than ones whose quality ratings either diminished or stayed the same. A study by Buzzell and Wiersema showed that gains of four percentage points were achieved when product quality was improved—not an insignificant amount.[9] This study seems to support the idea that perceived quality provides a utility to the consumer that influences the supply-demand relationship by more than cost alone. The quality-market relationship also applies to service industries, as Buzzell and Gale show.[10]

An example underscoring this important quality–market share relationship is the Xerox Corporation, a 1989 winner of the Malcolm Baldrige National Quality Award. A *New York Times* article describes how the company's focus on quality increased its market share:

> Xerox officials say this emphasis on quality and the attendant reduction of cost has made the company one of the few to regain market share from Japanese manufacturers without government assistance. Xerox's unit share of the market for smaller machines capable of making 12 to 30 copies a minute stood at 17 percent in 1979. After the Japanese entered, that share was pushed down to 8.6 percent. But by 1988, it had grown back to 15 percent.[11]

In their book *Delivering Quality Service*, Zeithaml, Parasuraman, and Berry point out other strategic advantages of service quality:

> Superior quality is proving to be a winning competitive strategy. McDonald's. Federal Express. Nordstrom. American Airlines. American Express. L.L. Bean. Domino's Pizza. Disney World. Club Med. Delux Corporation. Marriott. IBM. In every nook and cranny of the service economy, the leading companies are obsessed with service excellence. They use service to be different; they use service to increase productivity; they use service to fan positive word-of-mouth advertising; they use service to seek some shelter from price competition.[12]

Quality Improvement and TQM

Today's firm is the product of over a century of evolution. The picture of it that emerges is one of an enterprise organized into specialized functional entities by means of a command structure. Work flows across these vertically organized functions. Organizations, both profit and nonprofit, inherently contain the hidden plant phenomenon and elements of suboptimization and segmentation. These elements work against productivity and operational effectiveness and affect the quality of output. As a result, a typical organization today has substantial opportunity for improvement.

There is increasing realization on the part of management that quality has become an important factor in gaining market share and improving profitability. The question now becomes, What is the most effective way of addressing this crucial issue?

Over the years numerous approaches to quality improvement were developed. In the 1960s the zero-defect program was adopted by many manufacturing companies in the United States with little success. In several instances, this effort emphasized exhortation and publicity but was lacking in specific actions to improve product quality. In the 1970s prescriptive approaches emerged through the work of Phillip Crosby[13] and Joseph Juran.[14] W. Edwards Deming developed a multidictum approach consisting of 14 points.[15] Whereas Juran has promulgated a problem-specific (project) concept for improvement, the Crosby 14-step approach involves management commitment, organization (quality councils), cost of quality, problem resolution, and participant recognition. Deming provides a mix of caveats (e.g., don't buy based on cost alone) and prescriptive points (break down the barriers, employee training). Juran provides a structured approach to problem solving once the improvement project is selected. None of these, however, examines an organization in terms of the functioning of its productive system and the manner in which the system is organized.

In recent years generalized quality improvement activity has become known as total quality management, or TQM. TQM had its origins in the work of Armand Feigenbaum in the 1960s. In his book *Total Quality Control*, he points out that the quality of a product does not result solely from the manufacturing function. Other functions such as product development and field service also contribute to the quality of a product. The procurement function, through its efforts in obtaining suppliers for manufacturing, is involved in product quality as well. Feigenbaum's ideas were adopted and modified by the Japanese and called Total Quality Control (TQC) or companywide quality control (CWQC). The concept of TQC began to mean all functions of the firm cooperating in achieving customer-based product quality.

In the late 1970s and early 1980s, as the quality improvement effort in the United States began in earnest, total quality control was rediscovered as firms began to realize that product quality was not solely the responsibility of the manufacturing function or the quality control department. The TQC concept then became known as total quality improvement or total quality management and was given the acronym TQM. Today TQM is much broader in scope than was originally visualized as TQC. It involves not only activities that improve the product or service but also all the supporting activities of a firm, from secretarial services to groundskeeping and maintenance.

An approach that produces consistent and effective results, provides an integrated, unified approach toward TQM, and provides a sharp focus on the manner in which a business is conducted is process management. Process management is a concept that forces a focus on the flow of work independent of whether work is a product or service and independent of organization. The product or service output of a firm is the result of a series of work activities that comprise transformation of material and information. This set of work activities that produce an output is known as a *process*. Process management provides a method of analyzing an operation by directly addressing the organization, its nature of work, and how it is conducted.

The concept of process management was developed at IBM, where examining business operations as a process has become an important approach to total quality improvement.[16] The principles of process management have been applied to analyze and improve key processes of its business operations.[17] Functions such as finance, purchasing, product release,[18] marketing,[19] and product development[20] have applied this concept with excellent payback in terms of cost savings and operational effectiveness.

The first part of this book is devoted to examining the features and characteristics of a process, which form the foundation of process management. Following this the three phases of process management are described, after which techniques for analysis and criteria for evaluation are reviewed. At the end of Part I, a cohesion example is provided illustrating a case in applying process management and its role in TQM. Part II is devoted to describing various applications of the concept in staff, service, financial, and laboratory operations. Finally, in Part III criteria and considerations for designing a process are reviewed. The final chapter describes future directions that are developing in business operations.

With this perspective, let us now turn to the concept of a process and examine some of the important developments that provide a foundation for process thinking.

Notes

1. J. Main, "Battling Your Own Bureaucracy." *Fortune,* June 29, 1981.
2. M. Green and J. F. Berry, *The Challenge of Hidden Profits,* William Morrow, New York, 1987.
3. A. D. Chandler, Jr., *Strategy and Structure,* Anchor-Doubleday, New York, 1966.
4. A. Feigenbaum, "Quality and Business Growth Today," *Quality Progress,* November 1982.
5. L. Thurow, et al. "Interim Results of the MIT Commission on Industrial Productivity," AAAS Annual Meeting, Boston, February 15, 1988.
6. R. M. Kanter, *The Change Masters,* Simon & Schuster, New York, 1983.
7. J. B. Quinn and C. E. Gagnon, "Will Service Follow Manufacturing Into Decline?," *Harvard Business Review,* November–December 1986.
8. J. Ryan, "1987 ASQC/Gallup Survey," *Quality Progress,* December 1987.
9. R. D. Buzzell and F. D. Wiersema, "Successful Share-Building Strategies," *Harvard Business Review,* January–February 1981.
10. R. D. Buzzell and B. T. Gale, *The PIMS Principles,* Free Press, New York, 1987.
11. "Stress on Quality Raises Market Share at Xerox," *New York Times,* November 9, 1989. Reprinted by permission, New York Times Company, Copyright © 1989.
12. From V. A. Zeithaml, A. Parasuraman, and L.L. Berry, *Delivering Quality Service,* Free Press, 1990. Reprinted by permission of Macmillan Publishing Co., New York, Copyright © 1990.
13. P. Crosby, *Quality is Free,* McGraw-Hill, 1979.
14. J. E. Juran, *Managerial Breakthrough,* McGraw-Hill, New York, 1964.
15. W. E. Deming, *Quality, Productivity and Competitive Position,* Massachusetts Institute of Technology, Center for Advanced Engineering Study, 1982.
16. E. J. Kane, "IBM's Quality Focus on the Business Process," *Quality Progress,* April 1986.
17. E. H. Melan, "Process Management in Service and Administrative Operations," *Quality Progress,* June 1985.
18. J. Pazera, "Productivity Gains through QIBP," *Dimensions,* June–July 1987, IBM Corp., Kingston, New York.
19. W. A. Nickell and J. S. McNeil, "Process Management in a Marketing Environment," Proceedings, IMPRO 1986, Juran Institute, Wilton, Connecticut.
20. E. H. Melan, "Quality Improvement in an Engineering Laboratory," *Quality Progress,* June 1987; F. Mancuso, "Teamwork A Focus on Engineering Quality," Proceedings, 42nd Annual Quality Congress, May 1988.

2

Origins and Characteristics of a Process

Although it is only recently that service operations have been viewed as a process, the notion of a process is deeply rooted in our Western industrial society and dates back to the early days of the Industrial Revolution. By 1800, the Soho Engineering Foundry in England was applying the concept of sequential tasks to manufacturing.[1] Detailed descriptions were provided of metal casting operations and the flow of work from beginning to end.

Charles Babbage, in his 1832 treatise *On the Economy of Machinery and Manufactures,* added to these early concepts. In addition to his emphasis on various aspects of division of labor, Babbage pointed out the importance of analysis of manufacturing processes to establish product cost. Of key importance was his understanding of the advantages of combining and centralizing work activities—evidence of recognition that barriers to work flow existed even in the early days of manufacturing.

Seventy years later, Frederick W. Taylor, known as the father of scientific management, added to the growing body of process knowledge. Although more widely known for his pioneering work in time and motion studies, his espousal of analysis of sequential tasks is important from the viewpoint of understanding the origins of the process concept.[2] Taylor's work also provided the impetus for the field of industrial engineering and time and motion study methods.

The notions of work flow, sequential operations, and their control were further developed and refined for over half a century after the publication of Taylor's work and helped to create efficient industrial processes. Most of this work, however, was aimed directly at production activity; it remained for others working outside of the manufacturing domain to show applicability to service operations. It also remained for others to demonstrate the social and behavioral dimension of labor and the need to take these factors into account in understanding the nature of work.

Process Elements

From an operational viewpoint, a process is a bounded set of interrelated work activities each having prescribed inputs and outputs. It has a well-defined beginning and end. A process is essentially "a method for doing things."[3] The main purpose of a productive process is to create from a set of inputs one or more outputs of greater added value than the inputs.

There are three key elements common to any productive process:

transformation

feedback control

repeatability

Output is fundamentally the result of a transformation or set of transformations. In the model shown in Figure 2-1, inputs, whether they are material, equipment, other tangible objects, or various kinds of information, are converted by a series of activities into an output that is provided to a recipient. The recipient may be a revenue-paying customer, a department or group (in the case of an internal operation), other processing equipment and machinery, or the outputs may simply be stored for future use.

These transformations can be classified as:

- Physical
- Locational
- Transactional
- Informational

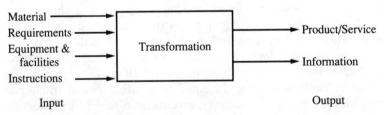

Figure 2-1. Transformation model.

A physical transformation involves converting raw or semifinished material, together with certain information relative to the conversion, into an output of higher added value. Physical transformation can be thought of as a change of state. A manufacturing process such as assembling an automobile is an example of this type of transformation. Locational transformations, on the other hand, involve movement of objects or material from one place to another, and storage as in moving and warehousing.* Transactional transformations involve a value-implied exchange, such as banking or stock brokerage. Informational transformations comprise data reduction and modification. Here, data are the input. The transformation process is modification of data and the output is data that have been converted into some meaningful information, such as a financial report. Data processing and financial planning are examples of this type of transformation.

A process typically contains at least one, and often two or more, of these basic transformations. Service processes (which include administrative and support types of activity) also include the main types of transformations described above. Many of them involve transactional, locational, and informational conversions. Banking and insurance, for example, are service processes which are largely transactional in nature. Transportation is basically locational with some transactional conversion occurring. Data processing and product development operations are primarily informational transformations. A transformation can be classified in terms of its primary conversion mode, as shown in Table 2-1.

A process, then, can be defined as a bounded group of interrelated work activities providing output of greater value than the inputs by means of one or more transformations.

Feedback control involves some regulatory means by which the transformation activities are modified or corrected to maintain certain attributes of the output. Any productive process requires feedback to regulate its output. Feedback can take the form of information of varying kinds from the output side of the process as well as information internal to the conversion processes. Feedback can also take the form of an economic return (such as revenue) which is used to sustain the operation. In general terms, feedback is needed to avoid degradation of the process.

---

*The physical distribution of goods from a manufacturer or distributor to a point of use can be viewed as creating added value. However, there are certain locational transformations (such as movement of parts within a manufacturer's warehouse) that have no added value and are, therefore, a waste. This is an underlying assumption behind the "just-in-time" (JIT) approach used widely in manufacturing today.

TABLE 2-1   Processes Classified by Transformation

Process/industry	Primary transformation mode
Banking, finance	Transactional
Construction	Physical
Data Processing	Informational
Health services	Physical (Physiological)
Insurance	Transactional
Manufacturing	Physical
Retailing	Transactional
Storage, warehousing	Locational
Transportation	Locational

Repeatability or cyclicality is the third key element. Repeatability implies that a process is executed many times in the same manner. Some processes are continuous in nature while other operate cyclically or intermittently. Chemical processes such as refining are continuous; a custom car works can be considered an intermittent process. Processes that involve groupings of physical items in the flow of work are known as batch processes.

Transformation Model

Let us now turn to the process itself and examine some of its features, using a generic process model. A process generally begins with inputs such as raw or semifinished material or data of various kinds. Certain prescribed transformations are performed at operation O_1 in Figure 2-2. Activities are defined by formal documentation such as process specifications, operational descriptions or routings, instructions of various kinds, drawings, and so on. The end result of the O_1 operation is a work product of higher added value that is then moved to a succeeding operation, O_2, where other activities are performed. The output progressively gains in value until it exits the last operation, O_n, as a fianl output or end product. In many physical and informational type processes, work flows from a position of lowest added value on the input side to a position of highest added value on the output side. However, there are exceptions in certain types of processes where the highest added value may occur early in the process.

Operation O_2' may be thought of as a control and measurement point where the work is sampled and statistical process controls may be applied. O_3' may be a form of a "re-do" operation such as rework or reconciliation, and O_n may be a final verification or a shipping-staging-transfer step. Note that the generic process shown has well-defined boundaries (inputs to O_1, output at O_n). Boundaries delineate the input and output sides of the work flow domain. Internal to the proc-

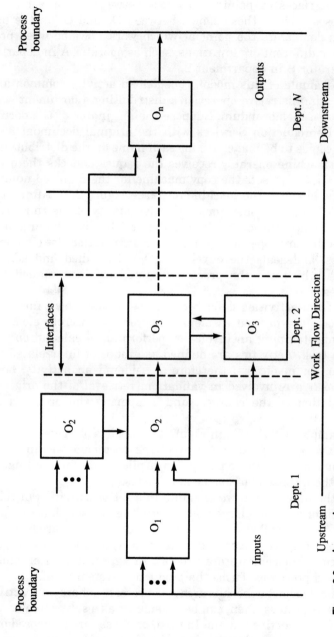

Figure 2-2. A generic process.

ess are interfaces or points of contact between major activities or subprocesses such as those shown between O_2 and O_3 in the figure. Interfaces demarcate the point at which work flows between process elements in different organizations, such as operator A in department 1 and operator B in department 2.

As an example of this model, consider an activity common among staff and administrative operations: distributing a document such as a report or a memorandum. Suppose the originator of the document provides Reproduction Services with the original document and requests N copies to be made, one for each name on the distribution list. The copy machine operator receives the request, checks the original for its quality (O_1), sets the copy machine for the required number of copies, and operates the machine (O_2) (see Figure 2-3). After completion of the work, the operator checks the quality of the run (O_3) and sends the completed job to the originator (O_4). The originator then addresses the envelopes according to the distribution list (O_5), inserts the copies, and seals the envelopes (O_6). The filled and addressed envelopes, which are the process outputs, are then mailed to the addressees.

Each of the activities described may be broken down further into tasks. Tasks are defined as elemental work or discrete elements of work. Time-stamping an invoice or performing a calculation is considered a task. Activities are defined as groups of interrelated tasks. Validating an invoice is, therefore, considered an activity, because several tasks are involved in validation: retrieval of the original order, validation of the charges, and comparing the order with the invoice.

For example, activity O_6 in Figure 2-3 consists of tasks such as taking an envelope, inserting a copy of the document, sealing the envelope, and putting the envelope on a pile ready for mailing. Each activity, then, consists of two or more tasks.

If, on the other hand, we examine this set of activities in relation to its position within a larger set of activities, we see that it is part of a bigger domain called a subprocess. So, document reproduction and distribution may be considered a subprocess, a subset of a larger set of activities that constitute a process. In aggregate, an ensemble of interrelated processes, forms the productive system itself.

Figure 2-4 depicts the hierarchy of process, subprocess, activities, and tasks. A process, then, can be considered as a subsystem or a component of a productive system. In turn, processes are composed of subprocesses, which are related groups of activities. Many processes today are complex, consisting of numerous subprocesses and crossing several functional entities of an organization. These are sometimes called

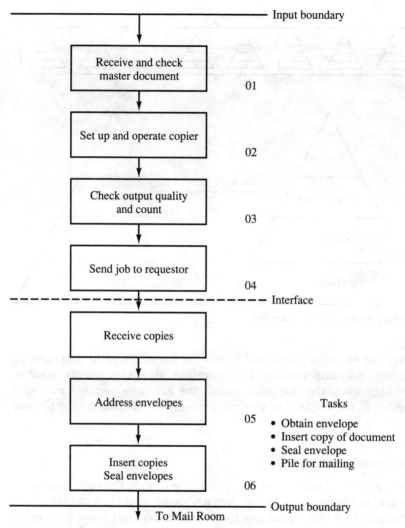

Figure 2-3. Document distribution process.

cross-functional processes. For example, order entry in a manufacturing firm may be considered a subprocess of its marketing process. Functions included in order entry not only involve the sales transaction itself but such other functions as information systems, production control, and accounts receivable. Order entry, therefore, is a cross-functional subprocess.

In practice, however, there is a tendency to call a subprocess a process, reflecting a reference point from which the specific operation is

System

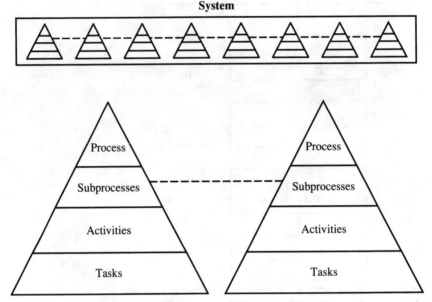

Figure 2-4.   System-process hierarchy.

viewed. For example, a controller may see his or her total operation as a process and components of the operation, such as accounts payable, as a subprocess. On the other hand, the accounts payable manager reporting to the controller may view his or her operation as a process rather than a subprocess.

Characteristics of a Process

Process management has been implicit in well-managed production operations, resulting in processes that are under real-time management control and providing products to the consumer that meet cost, quality, and volume requirements. Chemical, pharmaceutical, and high-technology operations such as semiconductors are examples. However, applying the same concepts to service and support activities has not been widespread, although the payoff in quality and productivity improvement can be substantial.

The principles of process management have their origin in the characteristics of a well-managed manufacturing process. To identify its general attributes and develop a foundation for process management, we now examine the features of a manufacturing process. From this, we compare these attributes to service processes.

A well-managed manufacturing process has the following character-istics:

1. *Clearly defined ownership.*   Traditionally, ownership of a manu-facturing operation is generally clear and explicit: it resides with a manager. The manager responsible for the operation is readily identifiable. This manager understands the organization mission, its output, and what he or she is accountable for. In addition, there are standards by which the manager's performance is judged, such as cost, schedule, and quality.

   In recent years, however, traditional management ownership has been gradually giving way to empowered work groups and self-directed work teams where employees are assuming some of the traditional roles of management.[4] A process owner, whether an in-dividual or a team, is fully responsible for yield, cost, quality, and schedule, and must manage the process to the targets set on these standards. Further, an owner has the authority to change or oversee a change in the process within his or her area of jurisdiction.

2. *Defined boundaries.*   Manufacturing processes have a clearly de-fined beginning and end. The final output, or deliverable, as well as the input(s) required to create it are clear and unambiguous. What is sometimes not clear, however, is whether output speci-fications truly reflect customer requirements and whether input specifications represent what is needed in the ensuing transfor-mations. The lack of understanding of requirements on either the input side or output side underlie many business processes. In a well-managed manufacturing process, requirements problems are minimized through conscious effort aimed at specifying the work product as it proceeds from one operation to another.

3. *Documented flow of work.*   Work flow in a manufacturing process is generally documented in great detail. There are several reasons for this. Documentation provides a permanent record of the man-ner in which a physical transformation takes place for production purposes. This record also provides a reference point or baseline from which any changes are to be made and serves as a means for replicating the process. Finally, documentation also serves as both a training and reference aid for the personnel involved in the process.

   Various types of documents exist; typical documents are process flow charts, assembly drawings, and routings. Flow charts provide a graphical description of the sequence of operations required to perform the transformations. These are described in Chapter 4. As-sembly drawings show how an item is put together. Descriptions of operation steps—often called routings—accompany the process flow chart.

4. *Established control points.* Control points serve as a means for regulating the quality of work. Because of natural variation that occurs in physical processes, control points are established to manage variation. These points involve such activities as inspection, verification of required characteristics, and disposition of discrepant material.

5. *Established measurements.* Measurements provide a statistical basis for controlling the flow of work and managing variation. In addition to verification of the conformance of the final work product to specifications, in-line measurements are inherent in any well-managed process. It has been found that depending on end-of-line verification is not only too late but can be costly in terms of scrap and rework. The quality of the final product is likely to suffer. Measurements serve as a factual basis for taking corrective action on variations that may occur. Statistical techniques such as the control chart serve as useful tools for managing variations in many operations of a repetitive nature.

6. *Control of process deviations.* In managed processes, corrective action is performed in a timely manner and from a statistical basis when an undesirable variation occurs. Feedback and regulation are the heart of process control and, without control, the process loses its capability of providing consistent output quality.

In comparison to well-run manufacturing operations, many service and administrative processes have ambiguous ownership, measurements are frequently nonexistent, and activities are neither defined nor controlled. In essence, many service processes are simply not managed from a process point of view. Ownership is often unclear as to who is the responsible individual; little or no documentation of the process exists; measurements, for the most part, are nonexistent; and control exists only in reaction to some extraordinary event.

Table 2-2 contrasts the basic features of a process between manufacturing and staff-service operations. In general, properly managed manufacturing processes are well-defined, documented, and controlled. Managing a process involves establishing ownership, establishing boundaries, defining the process and its control points, performing measurements using pertinent parameters that can be used to control the process, and taking corrective action on deviations.

TABLE 2-2   Manufacturing and Staff-Service Process Features

| Features | Type of operation | |
	Manufacturing	Staff-service
Ownership	Clearly defined	Amiguous/multiple ownership
Boundaries and interfaces	Defined	Often unclear
Process definition	Formally documented	Little/none or lacking integration
Control points	Established	Frequently nonexistent
Measurements	Established	Often nonexistent/ qualitative
Corrective action	Performed	Done reactively, if performed

From a management point of view, the most serious consequence of not managing a set of activities as a process is the inability to achieve control. Another consequence is an inability to understand its effectiveness. An end result is that the process functions in a reactive mode. Many business operations today operate reactively.

Other characteristics of a process include capacity, effectiveness, efficiency, and adaptability. Capacity is defined in terms of the output rate of the process. As an example, an insurance claims process may have a capacity of 100 validated and approved claims per day. Capacity is usually expressed in terms of design or theoretical capacity and effective capacity. Design capacity is stated without consideration of such matters as equipment reliability and maintenance and personnel factors such as learning curves, absenteeism, illness, and so on. These considerations, when allowed for, enable effective capacity to be calculated. Effectiveness, efficiency, and adaptability are characteristics used to evaluate process "goodness" or the inherent quality of the process. These will be described in Chapter 8.

Service Processes

Managers in service industries often ask how a manufacturing process has any relevance to a service process. The answer is that both manufacturing and service operations have common elements of transformation, feedback control, and repeatability. They also have characteristics similar to those described.

In addition to these elements and characteristics, most service processes have the following distinguishing features:

1. *A degree of customer contact.* Service processes range in degree of contact with the customer from virtually none (as in the case of food preparation) to complete contact (as in the case of medical treatment). In contrast, manufacturing operations have little or no contact with the customer. Services such as legal, retailing, hotels, and hospitals require more people per unit of service than more capital-intensive services such as banks and airlines.

2. *Intangibility.* Often, the service provided cannot be touched or felt by the customer but is experienced; hence, it is intangible. A plane flight is an example of this.

3. *Immediacy.* Service is often created and delivered at the point of customer contact—sometimes with customer participation. Medical examinations are a case in point.

4. *Non-accumulation.* For the most part, services are consumed as they are created. Consequently, services cannot be accumulated or inventoried, in contrast to products.

Although these features may affect how quality and productivity are measured and controlled, many aspects of service processes remain common with manufacturing.

Work Flow and Organizational Structure

Traditionally, processes encompass a series of interrelated work activities that are organized and grouped by function. Management has, for some time, recognized the advantages of grouping work on a functional basis either by skill or specialty or work activity. Grouping by function enables resources to be pooled among different work activities in the organization. It allows and promotes specialization as well as efficient management of similarly skilled personnel. Even in the newer management concepts of focused factories and group technology, organizations are functionally structured.

In spite of these advantages, however, functional structures have certain inherent weaknesses. In this type of organization, there is an emphasis on nurturing and sustaining skill specialties. This often detracts from attention to the basic objective of the function, namely, work output. Management and workers within the function tend to focus on their own means rather than the broader ends of the organization. This has been termed suboptimization and is the underlying cause of the hidden plant phenomenon noted in Chapter 1. This tendency can be attributed to one or more of the following conditions:

1. A reward system within the function that promotes values and behavior different from that of the organization. For instance, an employee's performance (and, therefore, rewards) may be based on quantity of output rather than the quality of the output. Quality may be the prime objective of the firm, but the employee receives cues from his or her supervisor that quantity is first priority.

2. Group behavior that may encourage a parochial—or "us versus them"—attitude towards outsiders. This behavior manifests itself in suspicion of, or antagonism toward, other groups involved in the flow of work.

3. Culture and behavior patterns existing within the function. One group may have an attitude of helpfulness, cooperation, and "can do"; in contrast, another may have nothing but hostility and negativism. The difference in work output between the two groups is usually marked. Familiar remarks heard in the workplace such as: "It's not my problem," "Let George do it," "It's not my responsibility," reflect a parochial view of work resulting from behavior patterns.

4. The degree of decentralization of an organization. Organizations that are highly decentralized often resemble sovereign domains in their conduct with one another and epitomize the phenomenon of suboptimization. Decentralized organizations are frequently noted for their factional strife and antagonism toward, and suspicion of, other groups. On the other hand, centralized organizations that are well-managed recognize the danger of suboptimization and work toward minimizing its effects.

A common source of difficulty in work flow within functional organizations is that processes may cross several departments or functional groups and may even extend beyond organizational lines to other geographical locations and business areas of an enterprise. In this context, work flow is horizontal in nature, flowing through functional entities organized in a vertical command structure. Coordination of overall work flow is difficult and sometimes nonexistent in these types of processes. As Mintzberg[5] notes, "The functional structure lacks a built-in mechanism for coordinating the work flow." Recognition of this problem gave rise to the concept of matrix or project management—a frequently used approach for coordinating complex project efforts distributed among various functional organizations. The development and production of defense weapons systems is a prime example of the application of project management.

The notion of work flowing horizontally across vertically arranged organization structures is important in process management and is discussed further in Chapter 3. Many of the difficulties encountered in work flow result because of work flowing across an organizational entity. At this crossing point, work is changed from an output to an input. The output is "handed off" to someone else in another organization. Frequently, the output is not really what the recipient would like it to be.

Another difficulty with a functional organization is that problem resolution may be difficult to accomplish. When a product deficiency problem occurs, for example, it is frequently difficult to determine whether this deficiency is due to a manufacturing quality problem or a design problem. Finger-pointing often occurs and no one will take responsibility for resolving it, especially when negative rewards are perceived by those in the organization associated with the problem. Task forces, blue ribbon committees, and other action groups exemplify typical methods used to identify and resolve product or service problems in functional organizations.

Thus far, we have noted the basic elements and characteristics of processes. Transformation, feedback control, and repeatability constitute these basic elements. A generic transformation model representing a set of activities was presented followed by a perspective of the task, activity, subprocess, process, and system hierarchy. The notion of work flowing horizontally across vertically structured organizations was also presented. The basic characteristics of a well-managed manufacturing process was contrasted with a service/administrative process. Typical functional organizations have inherent weaknesses, a key one being suboptimization with its resultant effect on quality and productivity. These characteristics now lead to the fundamentals of process management described in the following chapters.

Notes

1. C. S. George, Jr., *The History of Management Thought,* Prentice Hall, Englewood Cliffs, NJ, 1968.
2. F. W. Taylor, *Principles of Scientific Management,* Harper & Brothers, New York, 1919.
3. D. A. Garvin, *Managing Quality,* The Free Press, New York, 1988.
4. "At Monsanto, Teamwork Works," *New York Times,* June 25, 1991.
5. H. Mintzberg, *The Structuring of Organizations,* Prentice-Hall, Englewood Cliffs, NJ, 1979.

3

Fundamentals of Process Management: Process Initialization

In Chapter 2, we described several characteristics of a process that serve as a rationale for process management. The fundamentals of process management can best be understood by examining its three phases:

Phase 1: process initialization

Phase 2: process definition

Phase 3: process control

The first phase is described in this chapter. In the process initialization phase, ownership and boundary setting are determined in order to establish operational responsibility and the scope, or extent, of the process.

Process Ownership

In complex, interfunctional processes such as those that exist in many of today's business operations, process ownership is often ambiguous. Lack of accountability paralyzes action-taking to correct work-related problems, which makes for inefficiency, poor morale, and dissension among both employees and management. The common remark, "It's not my responsibility," is symptomatic of lack of ownership. Ownership ambiguity is largely a function of process complexity, the organization's structure, its culture, and the styles and attitude of management within. It is often found with complex processes in decentralized organizations where laissez-faire and uninterested management exist.

The following description (by an employee of a health care clinic) reflects the confusion in process ownership that can occur:

> Ownership of any service industry is very ambiguous. The health care clinic in question is privately run by a number of doctors and, therefore, there is no one superior ruling authority. But even more important, the process itself is characterized by a continued change of command within the process. The first part of the process is run by a receptionist and an assistant, the second part of the process is run by the doctor and the third part by the receptionist again.

The first operating precept of process management, then, is establishing *process ownership*. Ownership generally connotes possession. An owner of a factory, for example, possesses the assets of the facility. Managers are agents of the owner and are responsible for the proper utilization of these assets without having direct ownership. On the other hand, process ownership implies a responsibility and accountability for a set of operations and does not necessarily involve direct possession of resources such as equipment and technology.

As a working definition, we postulate that a process owner is accountable for the functioning and performance of a process and has the authority to make, or oversee the making of, a change to it. For example, in changing the terms of payment in an accounts receivable process, approval by management is required unless an employee is empowered* to do so. Hence, a manager is the process owner.

Defining ownership is of critical importance in process management for these reasons:

- To have clear accountability for correcting deficiencies and making improvements to an operation.

- To facilitate problem resolution and action taking.

- To resolve jurisdictional issues that may arise. This is especially true of cross-functional processes. In a sense, the owner serves as a tie-breaker.

- To provide authority to make changes. The process owner has the right to either make a change in the operation or oversee a change in order to improve it for cost, schedule, quality, productivity and design reasons.

---

*As noted in Chapter 2, ownership has begun to shift from management to empowered work teams in the more enlightened organizations. However, ownership of complex, interfunctional processes is likely to reside with management.

Establishing and assigning ownership is fundamental to managing processes. Unfortunately, there are no prescribed rules for resolving ownership issues; each must be resolved on an individual basis. In some situations where ambiguity exists, assumed or ad hoc ownership can be taken to resolve the issue. In other instances, a stakeholder may assume it because a perceived lack of ownership may affect the performance of an operation that affects the stakeholder. In still other cases, escalation of the issue to the appropriate management level for decision is the proper course of action.

McCabe, in describing business process methods, notes several key aspects of ownership:

> The identification of a process owner is critical to improving business processes. The process owner does not have to have line responsibility for all aspects of the process and, in the case of cross-functional processes, will not.... The owner can be selected as the manager with the most resource invested, feeling the most pain when things go wrong, affected the most, or doing the most work.
>
> The owner should be at a high enough level to see the process as part of a larger picture, influence policy affecting the process and commit a plan for improvement. The owner is accountable for the process from its beginning to its end.[1]

Because of wide variations in culture, values, and operating environment among organizations, there are no standards that govern assignment of ownership; rather, each process must be evaluated on an individual basis and the most appropriate person identified and designated. Determining process ownership is generally straightforward in simple or flat organizational structures. It is frequently more difficult in decentralized organizations where authority is dispersed and, incidentally, process duplication may be prevalent.

In less formal organizations, establishing ownership can also present difficulties, as Sayles points out:

> The individual manager does not have a clearly bounded job with neatly-defined authorities and responsibilities. Rather, he is placed in the middle of a system of relationships out of which he must fashion an organization that will accomplish his objectives. There is no "standard" interface; rather, the relationships differ, depending on the objectives and the position of other groups with whom he must achieve a working pattern of give and take.[2]

Ownership ambiguity is common in complex processes because of the numerous interfaces and cross-functional relationships that exist. Determining (or clarifying) jurisdiction is a nontrivial problem. In simple processes that are contained within a work group, department, or function, establishing ownership is generally a simple task. It becomes a matter of determining which manager has the greatest span of accountability for the series of operations involved and formally designating that manager as the owner. Examples of ownership are provided in the situation described in Chapter 9 and in the cases described in Part II.

Determining ownership is usually not an easy matter in organizations where processes traverse geographic locations, divisions or operating groups, or national boundaries. For these kinds of processes, decisions at higher management levels may be required to assign ownership. In a large, multinational company, for example, vice presidents were named as owners of specific business processes that crossed national boundaries. In practice, for cross-functional cases, operational ownership is assigned to the manager within the organization most affected by the process. An example of establishing ownership in this type of process is given by Kane:

> The owner is typically the manager most responsible for the results of the process. The owner of the billing process, for example, is the vice president of business systems, an executive directly responsible for most of the activities...This vice president holds a very high level position; however, he does not have responsibility for areas such as field marketing or manufacturing....Nevertheless, as process owner, the vice president is in a good position to cause changes to be made in these other areas. If a disagreement should arise about a change in one of the billing areas not reporting directly to the vice president, an escalation procedure would carry the matter to the next higher level of the corporation. At this point, the owner is the advocate for change in the process.[3]

Kane further states that, because of the scope of responsibility and authority given the process owner, escalation occurs infrequently. It should also be noted that the use of escalation is a function of the culture of the organization, and of management styles and personalites. In certain complex cases, it may be more appropriate to define boundaries first and then determine ownership.

Even when it is difficult to establish ownership of the entire process, one can always define a portion of it in which ownership is clear and can be assigned without encountering jurisdictional disputes. Here, the criterion for ownership is accountability for a particular subprocess. Thus, a process can be initialized by first establishing practical (i.e., manageable) boundaries and then determining the owner. The remaining parts of the process requiring ownership definition can then be resolved through negotiation by peer managers or escalated to a higher level of management.

It should be noted that establishing ownership facilitates but does not guarantee process improvement. Improvement also requires owner involvement. It is important that the change agent, be it an individual or a process team, create a linkage with the owner in order to implement process improvement. Stravinskas[4] notes the difficulty that a team can encounter without involving the process owner:

> Many teams were comfortable in analyzing a problem and making recommendations but did not feel they had the power and authority to change the process. Increased involvement by the process owner, at this point, would give the team credibility and speed implementation.

In addition, timely implementation of recommendations is also necessary to sustain productive output of process teams. The process owner should provide feedback to the team on the progress of implementation or the reasons why a recommendation cannot be implemented.

Boundaries and Interfaces

The second major step performed in the initialization phase is to define boundaries within which the process is to be defined and managed and identify key interfaces in the flow of work. Process boundaries define the limits of a set of activities. For example, an accounts payable process may start with receipt of an invoice (input) and end with the mailing of a check (output). An ordering process may begin with receipt of a customer order and end with the shipment of the requested item. Here, the input boundary denotes that the process is initiated by receiving an order. The end of the process occurs at the output boundary denoting shipment of the order. Figure 2-2 in Chapter 2 shows how boundaries may be designated on a flow chart.

Most analytic approaches require a boundary-setting step. In systems analysis, for example, it is important to identify at the outset the activities that are inside the system being analyzed and those that are outside. In process management, boundaries are intended to demarcate the input and output sides of the work-flow domain. Defining boundaries not only makes it easier to pinpoint ownership but is also helpful in locating the critical interfaces of the process. The input boundary denotes the external interface between the primary supplier(s) and the process. The output boundary designates the interface with the customer or receiver of the process output.

Internal interfaces are organization transition points contained within a bounded process that denote the point where the work output of one activity becomes an input to the next activity, as shown in Figure 3-1. It represents the point within the process at which the work product leaves one organization (such as a department, function, or business unit) and enters the next. Upon crossing the interface, the

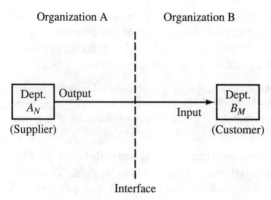

Figure 3-1.   A workflow interface between two departments or work groups.

work product becomes an input to the next step of the process. Interfaces also demarcate the point at which work flows between any two or more resource elements of the process, such as clerks A and B. For example, clerk A in a receiving department may time-stamp and log in an invoice, then send it to clerk B in the data-entry department for entry into the accounts receivable data base. Here A is the supplier (i.e., the producer) of the work product and B the receiver, or internal cusotmer.

Many problems relating to work flow originate at interfaces; hence, delineating interfaces becomes important in focusing attention on real or potential problem areas. Interface problems are frequently caused by lack of communication between the producer of the work and the receiver or customer on the requirements of the output.

Interface difficulties can seriously affect the quality of the output work product. An interface problem that resulted in disastrous social consequences was described in a *New York Times* article of May 9, 1986. It was the Chernobyl reactor accident in the USSR:

> Tucked into a long Izvestia report today on the Chernobyl reactor accident was a statement that "unfortunately" the service responsible for monitoring radiation inside the plant had had no contact with the service that monitored radiation outside the plant.
>
> Izvestia, the Soviet Government daily, did not further elaborate or explain. But the admission prompted the thought that, as radiation levels from the exposed reactor grew inside the stricken plant, monitors inside had given no warning to authorities outside.*

---

*Copyright © 1986 by the New York Times Company. Reprinted by permission.

Most readers would attribute this incident to a "communication problem." However, if we consider radiation monitoring as a subprocess within the total process of producing nuclear energy, it is clear that the subprocess either was not designed to allow communication flow between the internal and external operations (which appears unlikely) or that it was required but did not occur. Assuming the latter to be true, it is clearly an interface failure.

Another common cause of interface failures is a lack of understanding between the producer of the output and the receiver in terms of what is needed. In this example, a cross-functional sub-process—radiation monitoring—was evident, but no interface in fact existed. The cost to society of this failure is incalculable.

Customer-Producer-Supplier Model

A useful and effective way to address work flow inhibitors and improve a process is the customer-producer-supplier (CPS) model, Figure 3-2, which is based on the premise that a producer's work output must aim to satisfy the recipient's requirements. The recipient is the customer. A customer is the one receiving the work product and therefore may be internal to the process or external to it. The concept of internal and external customers is attributed to K. Ishikawa, a Japanese teacher and developer of quality management methods.

In the CPS model, there are three agents: the customer or receiver of the output, the producer or creator of the value-add output, and the supplier, who provides inputs to the producer. The model is based on three sequential phases:

- Output requirement phase
- Production capability phase
- Input requirement phase

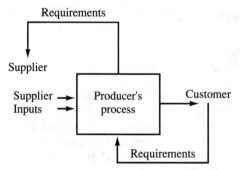

Figure 3-2. Customer–producer–supplier relationship.

Fundamental to this concept is that requirements have been mutually agreed to—often after negotiation between the producer and the customer, and the supplier and the producer. It makes no difference whether the output is for a "customer" internal to the operation that receives the producer's work product or for an external, revenue-paying customer—the requirements concept applies to both. In the case of accounts payable, it is as important for the data-entry clerk and the log-in clerk to agree on data requirements as it is for the issuer of the vendor's check to ascertain the accuracy and promptness requirements of the vendor (customer). In many instances, a customer may also play the role of a supplier. Here, a customer may not only be the recipient of the work product but may supply certain inputs to the producer (such as data, specifications, or documents). Again, an agreement between the supplier and producer should exist.

Output requirement phase

In the CPS model, one begins by defining the requirements for the final output of the process or subprocess.

In business operations, when work crosses either an interface between two operations or the output boundary it is accepted by the receiver of the work product. This acceptance is usually implicit. Acceptance of work implies that the work product received is in accordance with either expectation or specifications. In accounts payable, the data-entry clerk assumes that the document received from the log-in clerk is correct and has required, valid data. When the work product is deficient for some reason, one of three situations occurs:

- Work is accepted by the customer as is—with or without communication between the customer and producer as to its acceptability. The work is used without modification. Work that arrives late is typical of this situation.

- Work is rejected by the customer and returned to the producer for corrective action.

- Work is accepted by the customer and modified with or without communication between the customer and producer.

More often than not, it is because of a lack of understanding of customers' requirements that rejection and modification of work occurs. Hence, it is important to establish a clear, complete understanding of the characteristics of the work product between the producer and customer. This understanding should be documented to avoid future misunderstanding and misinterpretation of requirements.

Requirements documentation methods

The types of documentation used in establishing requirements are word descriptions, specifications, attribute lists, information flow charts, and deployment matrices. Any one or a combination of these may be used to define requirements.

Word descriptions.   Word descriptions, as the term implies, can be qualitative or quantitative descriptions of customer requirements. A simple qualitative word description that does this is the following: "The service provided shall be timely, courteous, and cost-competitive. It shall also be consistent and responsive as well as error-free."

Qualitative descriptions may form the source for attribute lists. They may also serve as product or service objectives. Word descriptions should convey a clear, unambiguous picture of customer requirements and should serve as a basis for more detailed documentation used in the transformation process.

Specifications.   Specifications, on the other hand, are both graphic (in terms of drawings) and quantitative descriptions of product or service attributes. These are generally of a technical nature and usually involve numerical values and tolerances. The size of an object, for example, could be designated in a specification. The phrase "error-free" is essentially a quantitative description of an attribute and is therefore a specification.

Attribute lists.   An attribute list is simply a list of properties or characteristics that the product or service must satisfy. In many service-type operations, attribute lists or word descriptions may suffice because many of the requirements do not have numerical characteristics. Consumer research, for example, may result in a product attribute stated as "low cost" or "economical." Frequently, however, an attribute is too general to be of use to a product developer or a producer because "low cost" can be interpreted in numerous ways. Often, an upper limit is placed on cost, allowing the producer to set cost targets for manufacturing the item. It is frequently necessary to ascribe numerical values to attributes desired by the customer in order to satisfy output requirements. Attribute lists can also serve as a basis for developing a deployment matrix (described below).

Information flow charts.   Information flow charts were originally developed in data processing to describe in graphical terms the flow of data. Requirements such as frequency and type of data input and length of data fields need to be specified in an exact fashion in order for software

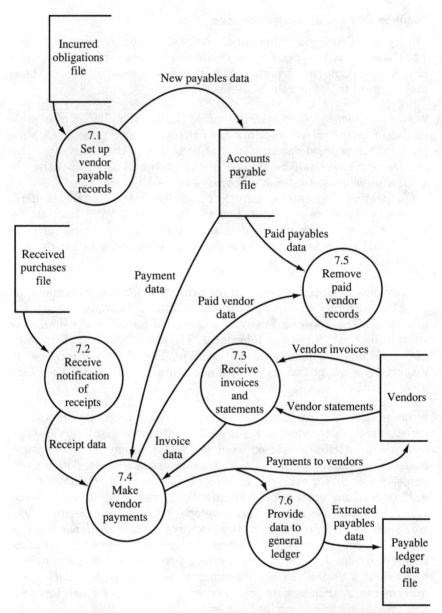

Figure 3-3.   Information flow diagram: Accounts payable. (Source: "Management Information Systems," R. McLeod, Jr., 3rd edition, copyright ©1986, Science Research Associates, Inc. Reprinted by permission of Macmillan Publishing Company, New York.)

to be developed for an application. A typical information flow diagram is shown in Figure 3-3. As can be seen, this type of diagram is based on the transformation model.

Deployment matrices. Deployment matrices represent a structured method for translating customer requirements (or wants) into characteristics of a product or service. These characteristics can be described in quantitative terms. As an example, if a customer survey resulted in the attribute "nonbreakable" for a flashlight, the designers of the flashlight, in order to use this requirement, may ascribe a certain impact strength to the housing. A numerical value of impact strength may dictate the use of certain materials for the housing. The initial deployment matrix will show a product attribute called nonbreakable on the left side of the matrix. A derivative characteristic reflecting a design parameter will have impact strength listed with a numerical value in the upper portion of the matrix. A symbol or a numerical value reflecting the relationship between an attribute and a characteristic is placed in the cell representing the conjoint relationship of the two. Typical relationships used are strong, medium, or weak. Scanning the symbols enables the viewer of the matrix to determine levels of importance in translating requirements into product or service attributes. Also, providing numerical weighting factors enables a quantitative measure of importance of each design attribute. The weighting factor method is described in Chapter 8 and shown in Table 8-3.

An example of a deployment matrix developed for a general service operation is shown in Figure 3-4. This matrix is set up in the following way. First, the customer-perceived requirements (or expectations) of good service are determined, employing one or more techniques (such as surveys, questionnaires, personal or group interviews, and focus groups). These form the elements of the lefthand side of the matrix, which is reserved for customer requirements.

Traditional deployment matrix methodology as described by Akao[5] allocates three levels of definition to required quality. The first level is generally stated as an abstract attribute. In this case, good service is a first-level quality characteristic or requirement to be met. The second level is a somewhat more specific derivative of the first-level attribute. Examining the results of the customer survey, we see that courtesy, attentiveness, and appearance can be combined into a more concrete, second-level characteristic called service attitude. Note that we decide to include personal appearance as an attribute composing an overall attitude or perception of service. Finally, third-level characteristics become courtesy, attentiveness, and appearance, which are specific attributes from which the second dimension of the matrix,

Substitute Quality Characteristics

1st level			Customer Satisfaction					
2nd level			Customer Service					
3rd level			Friendliness	Clean uniforms	Personal appearance	Attentiveness	Service manners	
Required Quality								
1st level	2nd level	3rd level						
Good Service	Service Attitude	Courtesy	N	W	W	S	S	
		Attentiveness	N	W	W	S	S	
		Appearance	—	S	S	W	W	

Figure 3-4. Deployment matrix—partial.

known as substitute quality characteristics, can be developed. The remaining characteristics can be apportioned among other second-level characteristics as needed. These third-level requirements are then prioritized by assigning a weighting factor to each requirement. It is not mandatory that customer requirements be developed to three levels. It is important, however, to state requirement attributes to a level specific enough to enable the detailing of a product or service design.

The second dimension of the matrix shown in the upper part of Figure 3-4 is reserved for specific and measurable elements derived from the third-level requirements listed at the left. As in the requirements portion of the matrix, the element section may also contain three definition levels. We can ascribe a first-level quality category called customer satisfaction that corresponds to good service. A second-level element called customer service is then designated; it is composed of the following third-level elements: friendliness, clean uniforms, personal appearance, attentiveness, and service manners. These elements are, by definition, quantifiable and measurable in some way. The resultant matrix, then, is a detailed map of required attributes and associated characteristics that are used in the design of the product or service. The problem now becomes one of distinguishing the importance of these characteristics. This is done by establishing relationship levels.

Once having developed the various attributes of the required quality and specified the corresponding quality elements in the second part of the matrix, an importance relationship is assigned to the cells. Each cell represents a conjoint third-level attribute and quality element. Relationship levels are ascribed to each attribute-element cell of the matrix in terms of strong, normal, and weak relationships. For example, courtesy and friendliness may be assigned a normal relationship, whereas appearance and clean uniforms have a strong relationship. Similarly, a strong relationship would exist between the appearance attribute and the element personal appearance. On the other hand, comparing appearance with service manners might result in a weak relationship. In this manner, every cell in the matrix is ascribed a relationship strength, or correlation.

Relationship strength can be designated by symbols, letters, or numerical values. The relationship levels assigned enable one to distinguish a degree of importance of the attribute as well as the degree of association among them. Relationship levels are often subjectively determined and, therefore, may vary from one individual to another. In Figure 3-4 the letters N, S, and W are used to designate normal, weak, and strong relationships respectively.

Deployment matrices are the basis for the quality function deployment technique.[6] QFD is a structured approach for delineating and translating requirements from a customer level to the most detailed technical level needed for designing and producing a product or service.[7] It reduces errors and omissions that frequently occur in translating customer requirements into technical characteristics. The literature is replete with cases of cost and design cycle time savings that have resulted from applying QFD. King[8] provides some of these examples and discusses some of the difficulties with the practical use of QFD.*

Deployment matrices are primarily used for transforming general requirements into specific attributes that can then be used for designing a product or service. The quality elements portion of the matrix enables measurements of the end item for control purposes. QFD matrices are often extended to include a competitive analysis section where the major quality attributes of the product or service are evaluated for each competitor. A description of the use and application of various types of matrices may be found in Brassard (see ref. 9).

Of the five basic types of documentation, word descriptions, specifications, and attribute lists are the most frequently used today to define

---

*King reports problems such as matrices becoming too large and, therefore, unwieldy; requirements being difficult to learn and categorize; and difficulty in determining the degree of relationship between required quality and substitute quality characteristics.

customer requirements. Of the five, however, only deployment matrices provide a structured approach to translating general, and oftentimes vague, requirements to an operational level where the requirement can be specified in terms of technical attributes or characteristics.

Once having documented preliminary requirements, the producer enters the production capability phase where requirements are examined in terms of their feasibility.

Production capability phase

In the production capability phase, the producer evaluates the capability of the process in meeting customer requirements. Frequently, the producer may conclude that the process as it exists is either unable to meet these requirements or that the inputs to the process constrain the output. The producer will either have to revert back to the first phase and negotiate with the customer to arrive at an acceptable set of requirements, or place new input requirements on the suppliers, or modify the process.

For instance, suppose Smith, in a firm's financial department, is required to issue a budget report on the first of every month. Smith, the producer in this case, requires various data from people (the suppliers) in five different departments one week prior to the due date. Understandably, Smith encounters difficulties in obtaining this data in a timely and accurate manner. Inputs are often late and incomplete. Smith must either renegotiate the output due date with his customer, modify the due date requirement placed on the suppliers, or change the process to accommodate shorter lead times.

Production capability may be assessed in terms of process attributes such as capacity, costs, quality, and responsiveness. These attributes are discussed further in Chapter 5.

Input requirement phase

Once the capability phase is complete, it is incumbent on the producer to negotiate with the suppliers of the inputs a set of requirements to satisfy the producer's needs. The supplier phase involves reaching an understanding regarding the requirements of the input with each supplier of input. Requirements may involve cost, timeliness, quantity, and quality characteristics. In many instances, particularly with internal operations, the customer also becomes a supplier. For example, design specifications or information of various kinds may be provided by the customer in order to achieve the required output. The types of documentation used to specify customer requirements are also applicable here.

Managing the Interface: A Matrix Method

In practice, the customer-producer-supplier relationship consists of a complex web of interrelated requirements. A technique that is useful in minimizing the boundary and interface problems that can exist in this relationship is the requirements-responsibility matrix. Here, all output or deliverable items are placed on one side of the matrix and the organization supplying the item is shown on the other (Figure 3-5). Each cell of the matrix represents a requirement as well as a commitment between the producer and supplier regarding this requirement. A cell represents either a specific deliverable or an understanding as to what is to be provided. In this figure, R1 could be the financial inputs required by Smith in the preceding example. F1 represents the first department providing this input. Similarly, R2 can be the same type of input required of the second department, and so on. The matrix, then, becomes a summary chart of all of the requirements to be satisfied by the suppliers shown in the matrix.

A simple illustration of a completed matrix for a cost analysis is shown in Figure 3-6. The X designation in the cell formed by the row labeled "indirect labor hours" and column D34 indicates that someone in D34 is responsible for providing indirect hour data for the department to the accounting department. Other X designations reflect a similar data provision responsibility to the accounting department.

More complex responsibility charts may have, instead of a simple X designation in a cell, a symbol or a number to designate specific requirements such as approval, consultation, information or awareness, and direct (line) responsibility. A completed chart, then, will reflect all possible types of interfaces and their requirements in conducting an operation.

Requirement	Responsible Organizations/Functions				
	F1	F2	F3	...	Fm
R1					
R2					
R3 ⋮ Rn		X			

Figure 3-5.   Requirements-responsibility matrix format.
X represents requirement R3 to be provided by function F2.

Requirement	Department responsible				
	D 22	D 34	D 73	D 98	D 65
Indirect labor hours		X	X	X	X
Direct labor hours	X	X		X	X
Absent hours	X	X		X	X
Sickness hours	X	X		X	X
Scheduled maintenance hours	X			X	
Unscheduled maintenance hours	X			X	
Computer usage hours			X		X

Figure 3-6.   Example of a requirements-responsibility matrix.

Matrices of this kind have been in existence for a number of years and have been found useful in managing complex projects. In general, a matrix technique is as useful a tool in process management as it is in project management, whether the process is simple or complex. It is sound management practice to define interface requirements and conditions. A matrix merely provides a tool for summarizing and tracking these requirements.

Figure 3-7 shows an example of an adaptation of a requirements-responsibility matrix used at a Ford Motor Company engine plant. Here, the work groups or functions participating in the process are indicated on both dimensions of the matrix and the requirement (operational objective) appears within the cell. The matrix now becomes a requirement involving the support groups and the customer. Hermann and Baker describe how the matrix approach, in combination with a producer-customer relationship model, was employed to obtain service commitments to support a manufacturing line:

> These proposed operational objectives were the subjects of give-and-take negotiation between the production department and service departments as well as between all of the departments. For example, production initially asked plant engineering, as an operational objective, to deliver 95% machine up time for a given line. To meet this request, which plant engineering might consider unusually high, production would have to provide 100% availability of machines for preventive maintenance...Since neither system was capable of this level of performance, a more reasonable operational objective had to be worked out by plant engineering and production.
>
> Similarly, in order to meet an operational objective of 90% machine up time, plant engineering would have to negotiate a "support objective" from the other service departments....For example, it might need a commitment from material control for a 100% supply of critical machine parts.

Operational Objectives		Support Objectives					
	Production	Quality Control	Plant Engrg.	Mfg. Engrg.	Material Control	Industrial Relations	Controller
"X" volume of engines per month — Production	■						
Quality Control		■					
90% Machine up time — Plant Engrg.	95% Availability of machine		■		100% Maintenance parts		
Mfg. Engrg.				■			
Material Control					■		
Industrial Relations						■	
Controller							■

Figure 3-7. The interlocking objectives matrix for a Ford engine plant. (Source: J. Hermann and E. M. Baker. "Teamwork Is Meeting Internal Customer Needs," *Quality Progress*, July 1985. Reprinted by permission.)

In round-robin fashion, each department—starting with production—played the role of customer until the matrix of interlocking objectives... was agreed upon. The first column contains the operational objectives of production. The remaining columns show the support objectives necessary for each department to achieve its operational objectives. Negotiations were facilitated by the plant manager and the internal consultant.[10]

Establishing process ownership and defining boundaries and interfaces comprise the initialization phase of process management. Properly defining the requirements that exist at the juncture between a customer and a producer as well as the producer and supplier is critical to the quality of the work flow. The CPS model is an effective concept for establishing product or service characteristics based on the interlocking relationship of customer, producer, and supplier. Establishing these features provides a sound foundation for the next phase of process management, namely, process definition. This phase is described in the next chapter.

Notes

1. W. J. McCabe, "Quality Methods Applied to the Business Process," Proceedings 40th Annual Quality Congress Transactions, 1986. Quotation reprinted by permission. © The American Society for Quality Control, Inc. 1986.
2. L. Sayles, *Managerial Behavior,* Krieger Publishing, Melbourne, FL, 1979. Quotation reprinted by permission of the author.
3. E. S. Kane, "IBM's Quality Focus on the Business Process," *Quality Progress,* April 1986.
4. J. Stravinskas, "Analysis of the Factors Impacting Quality Teams," 45th Annual Quality Congress Transactions, 1991. Quotation reprinted by permission, © The American Society for Quality Control, Inc. 1991.
5. Y. Akao, Ed., *Quality Deployment,* Goal/QPC, 1987, Methuen, MA.
6. L. P. Sullivan, "Quality Function Deployment," *Quality Progress,* June, 1986.
7. J. Hauser and D. Clausing, "The House of Quality," *Harvard Business Review,* May-June, 1988.
8. R. King, "Listening to the Voice of the Customer: Using the Quality Function Deployment System," *National Productivity Review,* Summer, 1987.
9. M. Brassard, *The Memory Jogger Plus +,* Goal/QPC, 1989, Methuen, MA.
10. J. Hermann and E. M. Baker, "Teamwork Is Meeting Internal Customer Needs," *Quality Progress,* July 1985.

Fundamentals of Process Management: Defining the Process

The second phase of process management involves defining the process. As noted in Chapter 2, many business processes, particularly service processes, are poorly defined or totally lacking in description. One frequently encounters work activity where procedures are simply by word of mouth or may reside in documents that are obsolete. Such activity is little understood by both participants and stakeholders. In situations like this, the remark "I'll never understand how it works" is often heard.

By defining the process, we provide a means for both understanding and communicating operational details to those involved. We also provide a baseline, or standard, for evaluating improvement. In many cases, merely defining the process as it exists can reveal glaring deficiencies such as redundant and needless steps and other non-value-add activities. Definition, then, becomes key to understanding an operation; in turn, understanding provides a basis for improvement.

Work activities and tasks can be described in words or by a combination of symbols and words. Word descriptions are often called procedures, or in the case of manufacturing activity, routings or operation sheets. Frequently, these documents comprise many pages of written detail and are difficult to visualize and understand. Symbolic descriptions are most often used in process definition and are called flow-process charts, process charts, process flowcharts, flow diagrams, and product process charts (PPC)—all equivalent terms. These are used to portray work and materials as well as data flow. The term *flowchart* will be used to describe these transformations. For some operations, word descriptions may accompany the flowchart.

Flowcharts and Symbols

Flowcharts are a graphic way to describe a group of transformations in productive systems. The basic purpose of these charts is to provide a symbolic representation of *all* the activities performed in the sequence in which they are actually conducted. The symbolic approach is generally more useful; process flow and the various activities can be readily visualized and understood. However, the level of detail that can be conveyed is less, in general, than pure word descriptions. Supplementary statements or footnotes sometimes accompany the flow chart to provide additional information.

Two types of flowchart symbols for analyzing any type of process are in popular use today. The first type is used in industrial and manufacturing engineering to describe unit operations* or transformation activities. Table 4-1 shows the main symbols used in industrial flowcharting. These geometric symbols have their origins in the symbols first developed by the Gilbreths in connection with work methods studies. The Gilbreths were prominent practitioners of the scientific management school. The American Society for Mechanical Engineers[1] and The Institute of Industrial Engineers have adopted five primary symbols for describing process flows: the circle designating an operation, the square or diamond for inspection or verification, the arrow for transportation or movement, the inverted triangle for storage, and the semicircle or capital D for delay or temporary storage. These symbols reflect five basic activities that can occur in a process:

- The work activity itself—typically a physical or informational transformation (the circle).

- A verification of the output of the transformation (the square or diamond).

- A movement of the output to either another transformation operation or storage (the line and arrow).

---

*Unit operation is a term used in describing chemical processes and refers to an activity that results in a change of state. Adding a catalyst to a chemical compound to cause a reaction would be considered a unit operation. The term was adopted later by researchers in the field of sociotechnical analysis.

TABLE 4-1   Product Process Flowchart Symbols

ASME Symbol	Name	Activity represented
(circle)	Transformation or modification operation	Modification or any change at all (changing shape or size, machining, permanent assembly or disassembly, etc.) of product at one workplace. Modification may be accomplished by machines and/or labor expenditure and does not necessarily add utilitarian value to the product
(arrow)	Move	Change in location of product from one workplace to another
		or
(D-shape)	Temporary storage/delay	Delay and waiting of work in process
(diamond)	Verification	Comparison of product with a standard of quantity or quality
(square)	Inspection	A control point established by management action
(triangle)	Storage	Store of raw or finished material

Adapted from G. Nadler, "Work Design" (Richard Irwin, Homewood, Il, 1970). Reprinted by permission.

- A storage of either work in process or completed work (the inverted triangle).
- A delay in the flow of work from either an operation, an inspection, or a movement (the semicircle).

In this traditional type of charting, one can readily see where inspections are performed, the various movements in the flow of work (as in Figure 4-1 part 2, for example) as well as delays. However, operations that involve decision-making and iterative activities such as rework, reconciliation (as in the case of accounting processes), or recycles (as in the case of design iterations) are not readily apparent when one views this type of chart. This disadvantage is avoided by using certain symbols developed for information flow charting, the second type of flowchart.

Two primary symbols are used in information flowcharting: the rectangle and the diamond. The rectangle, known as a process block, is employed here to describe a conversion activity. The diamond is used to denote a control activity of some kind such as an inspection where acceptance or rejection occurs. The line and arrow are used to indicate direction of the work flow. These symbols are shown in Table 4-2.

Two versions of a process flowchart for typing a document are illustrated in Figure 4-1. In part a, the work-flow for typing a document is depicted using the ASME symbols. The same activity is shown in part b using information flow symbols and demonstrates the graphic simplicity of the rectangle and diamond. However, certain information such as delay between activities is lost. This disadvantage can be overcome by time notation as shown in Figure 4-2.

Defining a Process

A step-by-step method for defining a process involves the following sequence:

1. First, establish the beginning and end of the set of activities or subprocess to be defined. By establishing a beginning and an end of an activity set, we have bounded or "scoped" it and can now describe all of the conversion and control activities contained within the boundaries.*

---

*Note that, in the process initialization phase described in the previous chapter, we bounded the total process to be studied. Here, we are bounding a set of work elements or subprocess within the overall process.

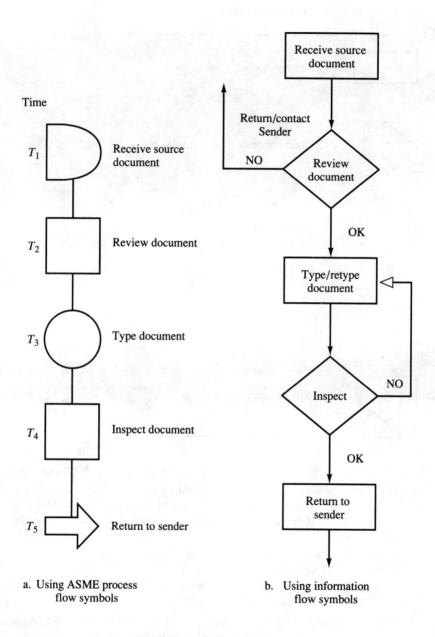

a. Using ASME process
   flow symbols

b. Using information
   flow symbols

Figure 4-1.   Process flowchart for typing a document.

TABLE 4-2 Symbols Used in Information Flowcharting

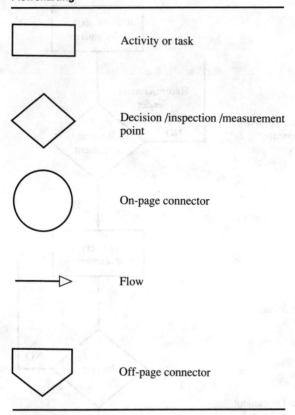

Activity or task

Decision /inspection /measurement point

On-page connector

Flow

Off-page connector

Every activity set begins with one or more inputs. These inputs are either material or informational in nature. Similarly, the end of this activity set is a work product output or outputs—the result of a set of transformations that have occurred. A form sheet such as that shown in Table 4-3 is helpful in process definition. On this sheet, the analyst can note the name of the activity set, the owner, the output(s) and the inputs to the set. Additional information such as number of interfaces and measurements (used or needed) should also be noted at this time.

2. Using whatever technique is appropriate for the situation—individual or group interviews of people working in the process, analyst observation, or review of pertinent documentation—a set of activities contained within the boundaries is flowcharted. Prior to flowcharting,

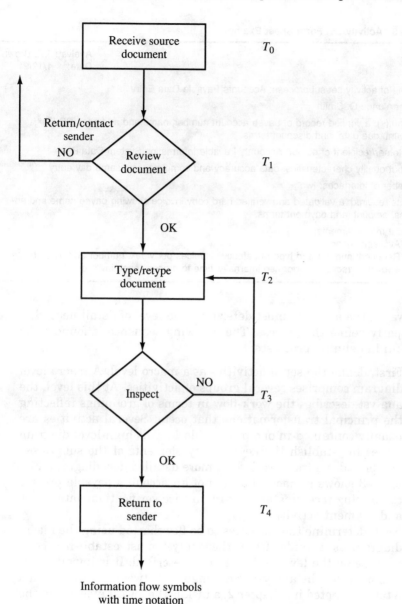

Information flow symbols
with time notation

Figure 4-2.   Information flow symbols with time notation.

TABLE 4-3   Activity Set Form Sheet Example

Analyst: J.X. Brown
Date: 1/12/91

1. Name of activity set/subprocess: Accounts Payable Data Entry

2. Owner: Jane Q. Smith

3. Output(s): a verified record of payee account number, name and address, IRS ID#, amount, due date, and discount terms.

4. Customer/recipient of output: Accounts Payable Information System Data Base

5. Output quality characteristics: data accuracy and completeness, same day entry

6. Number of interfaces: two

7. Inputs required: a validated and verified hard copy invoice showing payee name and address, amount, and payment terms.

8. Output measurements:
   a. Available: none
   b. Required: number and type of defective invoices per week, number and type of invoice corrections performed internally, time to process each invoice.

---

however, the analyst* must determine the level of detail needed to properly define the process. The following sequence is found to be useful in defining processes:

a. First, define the set of activities at a macro level. A macro level diagram comprises several groups of activities. At this level, the analyst describes the work-flow in terms of groupings reflecting the principal transformations that occur. Several activities are usually contained in one process block. This high-level diagram serves to establish the key activity elements of the subprocess and provides a framework for a more detailed flow diagram. Figure 4-3 shows a macro diagram of an accounts payable process comprising receiving and log-in, invoice verification, data entry, and payment activities.

b. Next, determine the activities to be flowcharted using the macro diagram as a guide. Here, the analyst must establish a compromise on the level of detail to be described. It is important at this point for the analyst to distinguish between an activity and a task. As noted in Chapter 2, a task is the lowest element in the

---

*It should be noted that an analyst does not have to be an expert in the process to be defined. The analyst, however, should understand how to describe a process symbolically by means of interviewing, observing, or extracting activity information from word descriptions.

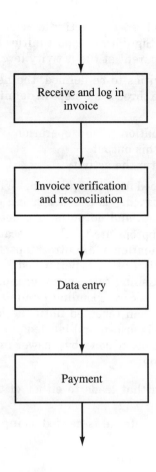

Figure 4-3. Macro definition of accounts payable.

process hierarchy. Time stamping an invoice in the case of accounts payable or inserting a bolt into a nut in the case of an assembly operation are termed tasks. However, a series of tasks such as time stamping, comparing the invoice with the original purchase order and entering the invoice data into an information system, would be considered the data entry activity portion of the accounts payable process.

Describing the operations at an activity level provides an adequate compromise between too little and too much detail. Insufficient detail provides little information on how the process operates or how to improve it; too much detail tends to cause confusion or may convey such an overwhelming amount of information that one may not know where to begin the analysis. On the other hand, once the activity level process chart is in hand, specific activities can be detailed at the task level if necessary. Significant opportunities for process improvement often become apparent at the activity level.

3. Having determined the group of activities to be defined, the next step is to develop a flow chart. There are three basic steps in defining each activity comprising a process:

   - Define the output for each transformation or unit operation
   - Define the work activity to provide this output
   - Define the inputs necessary to perform the activity

To determine which symbol is to be placed first, the question, "What is the very first activity that is performed?" is asked. Having determined this, the corresponding inputs and output(s) are determined and drawn on the chart with appropriate labeling. Because work flows through successive transformation and control steps, the next question is, "What is the next activity that is performed?" After having determined this and establishing corresponding inputs and outputs, the second activity with its corresponding inputs and outputs is then drawn. The steps are then repeated until the last activity adjacent to the exit boundary is drawn and labeled.

Determining inputs and outputs is often done by answering a series of questions such as the following:

- Does the activity involve a decision that leads to either of two output states?
- What work product, information, or material is needed as inputs to this activity?
- What are the requirements of these inputs?
- Does an interface exist between this activity and the preceding one?

Nadler suggests a simple approach for developing flowcharts:

> The greatest amount of information can be obtained by ... assuming that the person making the chart is one of the units being worked on. Then the question of what to record on the chart becomes "what happened to you as you progress through the sequence?" A symbol would be assigned to each thing done to you.[2]

Drawing Flowcharts

Flowcharts can be drawn either in a horizontal or vertical format, depending on individual preference or readability. If work is to be depicted as a horizontal flow, boundary lines can be drawn to the left of the first activity symbol and to the right of the last activity symbol as shown in Figure 4-4. Similarly, if work is to be described in a vertical flow (usually from top to bottom), a horizontal line can be drawn above the first activity symbol to designate the starting boundary. The ending boundary is then drawn below the last activity symbol.

For any process, complex or simple, it is beneficial to put in demarcation lines when work flow proceeds across organizational interfaces. These are shown as dashed lines in Figures 2-2 and 3-1. Providing interface lines enables the analyst or a process improvement team to focus on the output requirements as seen by the customer or recipient of the work product. As noted previously, interfaces tend to influence the effectiveness and efficiency of the process. Typical problems encountered in many administrative-service processes occur because of lack of communication between the producer and receiver of the work product. For this reason, it is important to maintain focus on the interface throughout the analysis.

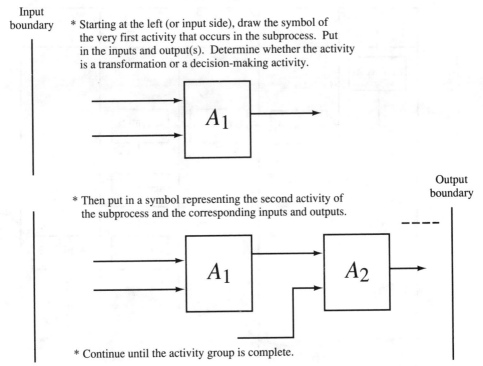

Input boundary

* Starting at the left (or input side), draw the symbol of the very first activity that occurs in the subprocess. Put in the inputs and output(s). Determine whether the activity is a transformation or a decision-making activity.

A_1

Output boundary

* Then put in a symbol representing the second activity of the subprocess and the corresponding inputs and outputs.

A_1 A_2

* Continue until the activity group is complete.

Figure 4-4.   Developing a flowchart.

For certain complex processes, it may be important to portray the different functions or departments involved in a set of operations, the activities within them, and the relationship of work-flow among the functions. The diagram in Figure 4-5 illustrates this type of flow chart. Boundaries and interfaces are clearly shown in this type of chart and, in fact, become an integral part of the work-flow symbology. This type of chart is also useful in showing how work flows back and forth between functional boundaries in less complex processes and may graphically demonstrate inefficient activity.

For the process analyst or the process team that is engaged in defining a process, revelation frequently occurs during the act of definition. Participants see redundant and other non-value-add activity and begin to question the value of keeping these activities. Self-discovery often provides motivation to improve the process. It has been the author's

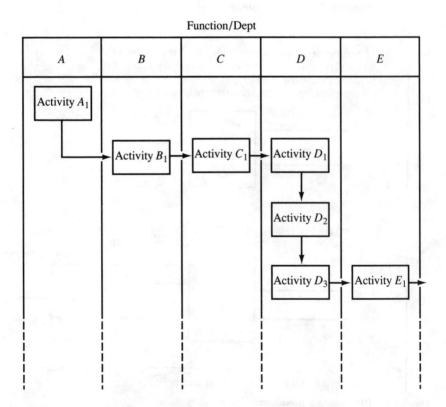

Figure 4-5.   Process flowchart for a cross-functional set of activities.

experience that defining a process at the activity level can provide months of improvement work. However, terminating process work after definition phase will never provide the process owner means for having the process achieve its full potential, nor will it provide the manager or team a basis for understanding its capability. For this, one needs to examine the third, and final, phase of process management: process control.

Notes

1. "Operation and Flow Process Charts," American Society of Mechanical Engineers, Cleveland, Ohio, 1947.
2. G. Nadler, *Work Design,* Richard Irwin, Homewood, IL, 1970.

Fundamentals of
Process Management:
Process Control

Having established ownership, set boundaries and interfaces, and defined the process, we reach the third and final phase of process management: process control. Process control consists of three steps: establishing points of control, implementing measurements, and regulating the process by obtaining feedback and performing corrective action.

There is a logical sequence to control. Before we can take corrective action, we must have measurements to serve as a basis for that action. In turn, for certain types of measurements to be made, control points must exist. Establishing points of control or determining their existence is, therefore, fundamental in managing a process.

Control Points

Control points are steps in the work flow associated with actions such as checking, inspection, auditing, physical measurement, or counting. A stage in the process that involves a verification or check of document quality, for example, can be considered a control point, because information obtained here may lead to interruption of the work flow or a modification of the preceding activity in order to improve its quality. Figure 5-1 illustrates a control point contained within a document processing set of activities.

Control is essential to any well-managed operation. In many business processes, operations that have no control points are lacking in internal feedback and regulation; control is based either on the as-

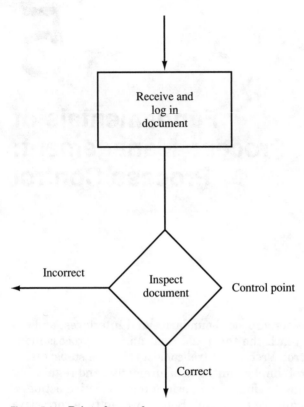

Figure 5-1.   Point of control.

sessment of the final output or on customer feedback to determine its quality (see Figure 5-2). When a process is operating in this manner, the quality of the output is not really known as it leaves the process. The process operates reactively—management relies mainly on feedback from the user to determine if product or service quality is acceptable. More often than not, the only feedback is negative—complaints that the product or service is not meeting expectation in one manner or another.

Reactive and Controlled Processes

In a reactive process, the quality of the final product exists prior to the output stage; little can be done to change its quality level other than by inspection and screening. Inspecting quality into a product is recognized as an inefficient and costly means of achieving quality.

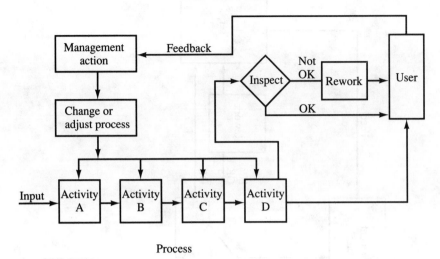

Process

Figure 5-2.   A reactive process.

When reliance is placed on customer feedback, discovery of poor quality is not only too late but becomes a greater risk to future business because customer dissatisfaction is a direct consequence. Correcting quality problems after the fact is expensive and impacts the image of the product and the reputation of the company. Field recalls and replacements are traditional actions taken by manufacturers for product type problems.

Service quality problems, on the other hand, are more difficult to solve. Because service is intangible and cannot be replaced, the only course of action is to correct the service process itself. Today, costs, business risks inherent in complex processes, and consumer attitudes all point to in-process control as the only method for producing quality products and services. With respect to services, the Strategic Planning Institute notes:

> Today, the principles of quality management are similar for product and service businesses. Traditionally, services were viewed as different. Since service is consumed as it is produced, final quality cannot be assured by inspection; process control is the only available method.[1]

Figure 5-3 illustrates the concept of process control. Instead of work proceeding to the final stage before examination, checks are made of the work in process and decisions made as to its quality. The points at which in-process checks are made are called control points. In-process checks often involve sample measurements. When these sample mea-

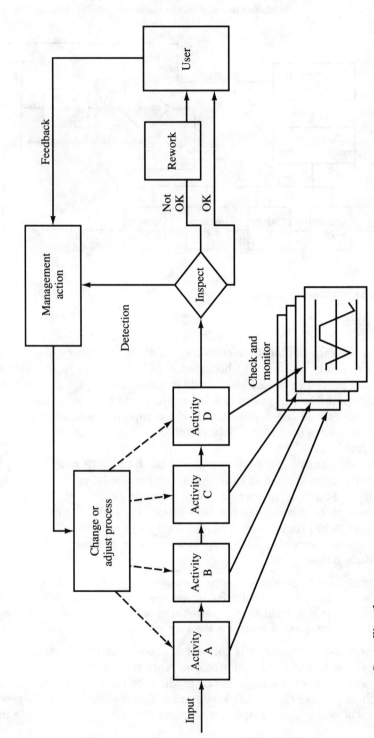

Figure 5-3. Controlling the process.

surements are plotted in time order sequence, we have the rudiments of statistical process control. In the case of physical transformation such as manufacturing a product, a sufficient number of in-process control points generally exist (or can be installed) so that measures can be developed for purposes of regulating the process. In low-contact service operations such as food preparation and handling and facilities management, in-process checks can be performed to assure that service quality standards are met.

Methods for Service Process Control

In the case of some high-contact services, however, it is difficult and, in some cases, impractical, to establish in-process points. Here, post-service checks such as comment forms, customer opinion surveys, and audits are the main methods used. Early evaluates some of these techniques:

> *Complaint and comment* forms can provide good initial indicators of problems. However, they are poor sources for data to diagnose and resolve problems, especially for some services. The greatest drawbacks for these data are that (1) they tell us nothing about the people who have never been customers and (2) they tell us nothing about those who were so dissatisfied that they took their business elsewhere without any comment whatsoever.
>
> *Direct customer surveys* are important tools in measuring quality if they are done right. Done poorly, they can be outright harmful.
>
> ...Since respondents may bias answers as the result of various kinds of conscious and unconscious clues, the best surveys are "double blind", that is, neither the respondent nor the interviewer knows for what company the survey is being conducted. Bias also comes as the result of any element of "self-selection" by the respondents. Once the sample is selected, reasonable efforts should be made to get responses from the selected sample. In that regard, the most expensive survey method is also the best - personal interviews are usually better than telephone surveys which are better than mail surveys. Cost and benefit must be traded off and one of the designs combining more than one method is often optimal.
>
> The content of the survey should capture four important pieces of information. First, what service features are important to the customer and what are their relative importance? These questions might well be answered in a separate, initial calibrating survey. But they must be answered by the customers. We cannot assume we know what is important. Once the identification and prioritization are complete, it may be possible, depending on the rapidity of change in the market, to keep them fixed for awhile. But failure to revalidate the identification and prioritization frequently enough could be a serious flaw.

The second critical element in the survey content is the identification of the customer expectations in terms of each of the service characteristics. These expectations must be specific. How soon is "on time"? How often is "frequent"?

Next, the customers and potential customers should be asked to evaluate our company in terms of their expectations. Finally, they should be asked to rank us against other suppliers of the same services.

Both what and how are important in services. The one factor may distinguish services more than anything else from manufacturing operations. The customer will react not only to whether technically good medical care was given in a hospital but also to whether the nurses, technicians and others were "friendly", "caring", "responsive", etc. The measurement of these human interactions requires not only customer surveys but also focus groups or other more intensive ways of exploring what factors enter into a perception of "friendly" or "caring."[2]

Audits such as sampling customer satisfaction or observing services being performed is a frequently employed control method. In a sample audit, the sample size can be statistically derived. The number of customers will be based on a confidence level, a sample error, and a preliminary estimate of the level or proportion of the attribute being studied. The equation that determines the sample size, n, is:

$$n = \frac{Z^2 pq}{h^2}$$

where Z is the standard normal deviate derived from the level of confidence desired for the audit, p is an estimate of the proportion of the attribute being studied (e.g., customer satisfaction), q is $1-p$, and h is half of the precision or accuracy interval. Since sample size is inversely proportional to the square of the precision interval, n can vary widely, depending on the accuracy required.

Values of Z for typical confidence intervals are shown in Table 5-1. As an example, suppose a service manager decides to do a sample audit of customers to determine the degree of satisfaction that exists

TABLE 5-1   Values of Z for Typical Confidence Intervals

Confidence interval (percent)	Z
80	1.28
90	1.65
95	1.96
99	2.58
99.7	3.00

with the services being provided by his operation. A preliminary estimate showed that 2 customers in 10 were dissatisfied, 1 was neutral, and the remaining were moderately or very satisfied. The manager decides that a 90 percent confidence and a precision of ±5 percent are reasonable. The audit sample size are then calculated:

$$n = \frac{(1.65)^2(0.7)(1 - 0.7)}{(0.05)^2} = \frac{0.5712}{0.0025} = 228$$

Hence, a random sample of 228 customers would be required to determine whether, in fact, a customer satisfaction of 70 percent ±5 percent exists. If the manager sees p below 65 percent or greater than 75 percent in the early phases of the audit, the sample size can be modified as the audit progresses.

In observing services being performed, the auditor may have a check sheet to note certain factors regarding the quality of the service (such as courtesy to the customer). In controlling customer service where transactions are performed by telephone (as in reservation systems), a supervisor randomly samples a number of employees daily, listening in on the conversations between the customer and the service provider. Here, a checksheet may be used for a quantitative determination of service quality. An example of a check sheet is shown in Figure 5-4. Note that certain of these quality characteristics (such as perceived customer satisfaction) are subjective and may therefore vary from one individual performing an audit to another. A typical solution to this type of problem is to provide training and assessment standards for auditing in order to reduce subjectivity.

Measurements

After establishing control points, determining the parameters to be measured is the next logical step in process management. Measurements are the heart of managing a process. They are needed not only to determine defect or error rates but to assess whether the output is conforming to requirements. Measurement may encompass simple sampling or auditing to complete inspection of every item. In business processes, the kinds of measurements that are performed fall into five categories:

Measures of conformance

Measures of response time

Measures of service level

NAME: Jane Smith   DATE: 6/15/91   SHIFT: 1

TOTAL CALLS MONITORED: 25

Characteristic	Tally	Number	Percent
Customer courtesy			
Positive	THL THL THL	15	60
Neutral/absent	THL I	6	24
Negative	I I I I	4	16
Responsiveness			
Above average	THL THL I I	12	48
Average	THL THL	10	40
Below average	I I I	3	12
Customer helped?			
Yes	THL THL THL I I I I	19	
No	THL I	6	24
Said "thank you"			
Yes	THL THL THL THL I I	22	88
No	I I I	3	12
Perceived customer satisfaction			
Very satisfied	THL I I I	8	32
Satisfied	THL THL I	11	44
Neutral	I I I I	4	16
Dissatisfied	I I	2	8

Figure 5-4.   Check sheet for customer service.

Measures of repetition

Measures of cost

Measures of Conformance

Conformance measures involve an inspection or verification as to whether the work product or service meets either a specification or some other requirement. Implicit in these measures is that they reflect customer requirements directly or indirectly. Clerical and source translation type errors such as miscodes are conformance errors that are measurable. The acceptable error rate may range from zero to some low percentage value, depending on the business or economic impact of these errors. An example of conformance measurements performed in a staff department is the following:

> The Technical Writing Services department reduced the defect rate in its technical publications by 97.3 percent in 16 months, beginning with a critical analysis of the entire writing and production process. The program called total defects per unit (TDPU) measures the ratio of defects detected in a 13-week period to the pages of technical literature completed. An average page of copy contains about 5,200 opportunities for defects. But, by using an elaborate quality screening process that makes the writer the monitor of defects, the average number of defects per page went down to about 2.00 per page within five weeks. At the end of 16 months, the defects per page had dropped to 0.54 or 14 defects per million.

This example resulted from work performed at Motorola, a 1989 Baldrige Award winner, as part of its Six Sigma Program.

As noted in Chapter 3, when output is in a state of nonconformance, one of three situations can occur:

- Work is accepted as is.
- Work is rejected and returned to the producer.
- Work is accepted by the customer and modified to conform to some desired state or condition.

These three conditions imply either that an ineffective inspection has been done on the work product or that no inspection is performed and nonconformity is detected solely through discovery in subsequent operations. Because of the consequences of nonconformance, verification controls are built into many processes.

Measures of Response Time

Response is measured from the arrival of the request until the completion of service or from the start of the actual performance of service until completion. For service facilities such as fast foods, package delivery, or the ten-minute oil change business, response time is crucial. It provides a competitive edge and serves to differentiate a firm from its competitors. Federal Express, for example, has emphasized nationwide overnight delivery service from its beginning in its attempt to differentiate itself from similar delivery services. Response or cycle time is generally an important measure for most service and support type operations.

In the case of product development, the cycle time in developing a product is critical to the competitive posture and market share of a firm, as noted in Chapter 13. Cycle time measures are also used in various staff and administrative functions. An example of this is the following:

> One comprehensive process being used at Motorola is cycle time management. While primarily a manufacturing term, cycle time can be applied to service tasks.... All administrative and staff functions at Motorola are working on cycle time reduction with success.
>
> The patent department, for example, has reduced the time it takes to file a patent from as much as two years to fewer than 90 days by getting patent attorneys involved with engineering, business and marketing people early in the process.[3]

General Electric has used cycle time measures to improve the delivery time of an order and reduce inventory costs. In GE's appliance group, the average production cycle for its consumer appliances was reduced 70 percent and the time between receiving an order and delivery for electric ranges and refrigerators was brought down to six days from sixteen weeks. These improvements have reduced inventories by greater than 50 percent and translate to annual savings of over $300 million by 1993.[4]

Measures of Service Level

The term "service level" means the degree to which a service or a facility is available to a user.* For example, an information system facility or data center might commit to provide on-line terminal availability to users 99 percent of the time during a month. This is known as a service-level agreement and is often embodied in a document. To determine that the service level agreement has been met, management would measure the percentage of time that the system is, in fact, available to the user during the month. In this case, the target value is 99 percent.

---

*The service level concept appears to have originated in inventory management where, for a particular inventory item, a stock-out risk is assigned. Service level is 1 − Stockout Risk. Thus, for a stockout risk of 5 percent, a service level of 0.95 exists. Depending on the type of inventory system and model used, an inventory reorder level is established for this item consistent with a service level of 95 percent.

Measures of Repetition

Measures of repetition involve measures of recurring events or the frequency of occurrence of an activity. Measuring the number of times a typed document is redone or the number of design iterations before the final design is reached are typical examples of a repetitive measure. Repetitive events are generally a reflection of wasted and unproductive work. Repeated rejection of a document that has not been typed properly adds significant cost to clerical activity.

Measures of Cost

Cost has always been a primary measure of business performance. However, the cost of product and service waste did not receive widespread attention by management until the advent of quality improvement in the United States. This cost concept known as "cost of quality" was promulgated in the 1950s by J. M. Juran[5] and A. V. Feigenbaum[6] and further publicized by P. Crosby.[7] Cost of quality has provided a powerful business approach for assessing the economic impact of waste in product manufacturing and service. For example, costs associated with scrap and rework of material, documentation errors, service and administrative errors of various kinds, as well as measures of loss such as stock-out and lost sales have been estimated to be in excess of 20 percent of sales for many companies. In large firms, quality costs can run in the tens of millions of dollars and even higher in some instances.

Quality costs are traditionally divided into three major elements:

1. *Failure or nonconformance costs.* Failure costs are those directly related to not meeting requirements. A nonconformance may be an incorrect insurance policy or a product not meeting specifications. Costs of product and material scrap and fixing defects or errors of various kinds are also failure costs. For an insurance policy, the cost of correcting a document having a wrongly calculated premium or an error in coverage would be considered a failure cost.

   Failure costs are often divided into costs of internal and external failure to enable management to focus on the appropriate areas of the business. An external failure cost would be that resulting from a warranty claim. An internal failure cost would be the scrap costs incurred because of faulty material used or generated in a production process.

2. *Appraisal costs.* Appraisal costs are those costs attributable to human and machine activity required to detect nonconformity in work. The cost of inspecting or auditing work as it proceeds from

one operation to another is an appraisal cost. The labor cost of checking an insurance policy for accuracy would be considered an appraisal cost.

3. *Prevention costs.*   Costs associated with preventing future occurrences of nonconformance are deemed prevention costs. The cost of developing a process improvement such as an automatic checking scheme for assessing the accuracy of a policy would be an example of prevention cost. Typical cost-of-quality assessments show that investments in prevention are small compared with appraisal and failure costs—a reflection of the short-term focus of business.

The quality cost approach, though by no means a standard, provides a comprehensive way to determine areas of an operation that are incurring waste and, therefore, require improvement. It can also show the savings in the failure and appraisal cost components when a preventive investment is made.

The more frequently used method in many business applications employs a straightforward computation of direct and indirect labor and equipment involved in an operation. Costs are based on the estimated cost of a nonconformance and any associated costs for repairing the error. In the case of response time, cost is based on labor time saved in producing a product or service.* Where time computations involve bringing a product to market earlier, estimates of the increase in sales are required in order to compute profits. For service-level agreement measures, the cost of lost sales (opportunity costs) or the cost of idle labor when a set of terminals is inoperative can be calculated.

Examples of various measures used in service and administrative operations are given in Table 5-2. These are not all–inclusive but are intended to provide the reader with an idea of the various types of measures that can be implemented.

Considerations in Selection and Implementation

Measurements must be meaningful, timely, accurate, and useful. Clearly, measures that are lacking in any one or more of these attributes will prevent proper control of an operation. In the selection

---

*Westinghouse, for example, uses a cost-time management concept for measuring process improvements. A cost-time profile is used in measuring an existing process; costs of materials, services, and equipment are depicted as vertical lines on a chart. Diagonal lines represent labor costs as a function of time spent on the process. Process improvements manifest themselves in reduced cost-time profiles and graphically show the reduction in both cost and time that results.

TABLE 5-2   Examples of Service/Administrative Measurements

1. Accounting and finance
   –Accounting report errors (by type and quantity)
   –Number of reports issued after due date (% per month)
   –Number of account reconciliations per month
   –Number of lost discounts per month
   –Payment error rate
   –Billing errors per month by error category
   –Number of defective vouchers per reporting period
   –Clock card or payroll transcription errors
   –Ledger reruns (time and frequency)
   –Deviation of predicted budgets from actual
   –Payroll errors (number per month by type)

2. Administrative
   –Employee turnover and transfer rates
   –Number of safety violations (per month by type)
   –Response time to inquiries
   –Typing errors (per month by type)
   –Keypunch or data entry errors (per month by type)
   –Time spend in locating filed material
   –Filing errors (per month or quarter)
   –Internal mail delivery time (average turnaround time by month)

3. Computer center services and programming
   –Computer outages (time and frequency)
   –Job turn around time
   –Number of program runs to a successful compile/test
   –Number of coding errors in inspection and test
   –Program error resolution time
   –Number of documentation errors

4. Engineering
   –Number of engineering change releases (per month by type)
   –Success rate in meeting product schedules
   –Response time to bid requests
   –Adherence to contract budget
   –Number of change requests due to design errors discovered in manufacturing and field
   –Design errors found in release packages

5. Field service
   –Number of repair callbacks (per month, by type)
   –Response time to service requests
   –Response time to spare parts demand
   –Time to repair (by type of repair)

6. Marketing
   –Customer inquiry response time
   –Order errors
   –Billing errors
   –Response time to special bids
   –Accuracy of market forecasts (predicted vs. actual)
   –Contract errors
   –Number of customers gained or lost per month
   –Number of late deliveries per week or month

and implementation of measurements, the following steps may be found helpful to the reader:

- First, determine what is to be controlled. It is useful at this point to determine (or review) the critical success factors of the process * and customer requirements. In many instances, response time is of primary importance. This makes a timing parameter such as cycle time or the elapsed time between the time an input is received until the output is transmitted an appropriate measure. In other cases, defect-free output is a critical requirement so a defect attribute such as errors becomes the right measure.

- Second, examine existing data bases for measures that are currently in use (if any). The question to be asked is: can what is available be used or can it be modified to extract the needed measures? This often occurs in the case of financial measures such as costs.

- If nothing is available, the question to be answered is: can a business case be made for new measurement tools, or should measurements be performed on a manual basis?

- Finally, given the type of measurements to be made, an appropriate sampling method, sample size, and measurement frequency must be determined. Here, the possibilities range from sample measurements of every item to sophisticated sampling techniques, depending on the application, intellectual understanding, skill, and statistical tools available.

A useful technique for detailing and summarizing the type and frequency of inspection is shown in Table 5-3. Here, the measurement parameter, its unit of measure, and frequency of measurement are listed for each activity. This listing serves as an accompaniment to the flow chart and becomes a central source of measurement information for the process.

Graphical Methods

Measurements of many business processes can be summarized by simple statistical charts such as trend charts and bar graphs (shown in Figure 5-5), pie charts and histograms, and even more complex charts

---

*The critical success factor concept was developed by N. Rockhart in connection with the application of information systems.[8] The concept is useful in examining any business process in terms of its critical success factors—business parameters that are crucial for the success and survival of an operation. For a firm, a critical success factor may be market share. For the marketing process, however, a critical success factor may be accurate assessment of the market for a product.

TABLE 5-3   Process Measurement

Process step (activity name)	Control parameter(s)	Unit of measure	Frequency	Remarks
A_1 Document inspection	Document defects – Missing – Illegible – Out of sequence	Percent of total errors by category	Every job	100% inspection
A_2				
A_3				
:				
A_N				

such as box plots. Some business processes are amenable to control chart techniques as Dmytrow[9] and McCabe[10] have shown. McCabe has also given several examples of administrative types of control charts for mail delivery and purchase order errors. These examples are shown in Figures 5-6 and 5-7.

Target Ratcheting and Improvement

In the case of trend charts, one may consider establishing sliding targets to reflect a continuous drive to improve the process as shown in Figure 5-5. The technique of "ratcheting" targets has been found to be an effective management technique to facilitate improvements. Target ratcheting involves the periodic reassessment of actual defect data in comparison with its target or objective value. When the actuals are consistently below target for a period of time, the target is adjusted to the actuals or below. In this manner, new goals are set periodically that support a philosophy of continual improvement. Targets may be derived from process capability studies, benchmarking, or traditional competitive comparisons.

An example of target ratcheting is IBM. In the 1980s, management of the IBM Poughkeepsie plant (which produces computer mainframes) began to hold manufacturing meetings with its counterparts in Montpelier, France and Yasu, Japan. These locations were considered "sister" plants in that the same type of product was being made at the same time by similar manufacturing processes. At these meetings, which were held quarterly, various topics such as process yields, product and process quality, process improvements, and costs were reviewed and information exchanged. During the quality reviews, each location would present its in-line performance data describing defects being

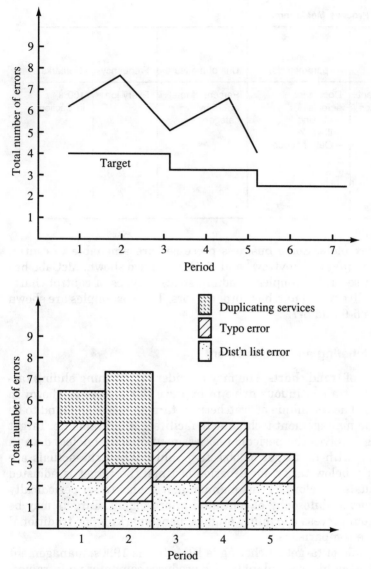

Figure 5-5.   Measuring the process.

encountered in its manufacturing process, its root cause analysis of the defects, and actions taken to prevent their recurrence.

The performance of each process sector would be presented in graphic form by a trend chart showing defects versus time. Each sector would have an established target, which was set both by past history and by judgement as to what the process is capable of giving. As the

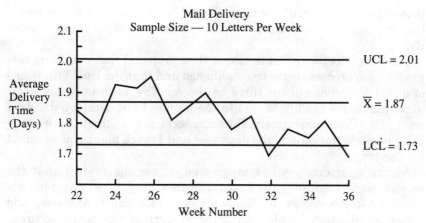

Figure 5-6. Example of service measurement control chart (variables).
(Source: W. J. McCabe, "Quality Methods Applied to the Business Process," Transactions 40th Annual Quality Congress, 1986; ©American Society for Quality Control, 1986; Reprinted by permission.)

quarterly reviews progressed, it became clear to the Poughkeepsie plant manager who attended each review that each manufacturing location had its own quality target. Moreover, for essentially the same process, each sector was achieving a different level of quality performance. In addition, it became obvious that no one plant was consistently superior to the other two for any period of time so that a detailed

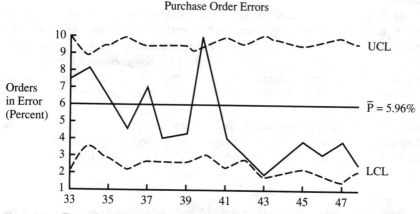

Figure 5-7. Example of service measurement attribute control chart.
(Source: W. J. McCabe, "Quality Methods Applied to the Business Process," Transactions 40th Annual Quality Congress, 1986; ©American Society for Quality Control, 1986; Reprinted by permission.)

root cause analysis could be made. It was also apparent that any learning curve differences could not explain these marked contrasts in quality.

The Poughkeepsie plant manager then proposed that a uniform target for each process sector be established and that the lowest (in terms of defect levels) one of the three be chosen. The Yasu and Montpelier plant managers readily agreed to the proposal and a target standard went into effect immediately. For some sectors, the targets were set by Poughkeepsie; for others, the Japanese and French plant management set the target.

As the quarterly reviews progressed, it became evident that the sectors were beginning to reach the new targets as the plan-do-check-act methodology of measurement, root cause analysis, and defect elimination became an integral part of the manufacturing process.* The rates of quality improvement differed by sector, by plant location and, to compound matters, differed in time. During the course of a year, the plant having the highest overall quality level shifted in what appeared to be a random fashion among Montpelier, Poughkeepsie, and Yasu. In fact, what was occurring was an interplant quality competition. As Poughkeepsie personnel saw that its defect levels (measured in parts per million) were somewhat greater than either or both of the other two plants, increased motivation to improve quality occurred. This, in turn, unleashed additional improvement efforts which resulted in further defect elimination and increased quality levels surpassing that of the sister plants. When this occurred, employees of the sister plants were goaded into further action to reduce defects and the cycle repeated itself.

At a point in time, all three plant locations were approaching targets. Each plant manager was faced with a dilemma: maintain the targets and be satisfied with their achievement, or change them. Each of the three knew about, and subscribed to, the concept of continuous improvement, hence, change was the only alternative. The question became how to do it? At the next quarterly review meeting, the Poughkeepsie plant manager proposed that, as each target was reached, a new target be automatically set to the actuals being achieved by the first location to break the target value. No location would be permitted to reset to a higher defect level. The Yasu and Montpelier plant

---

*Interestingly, this problem-solving approach was used without knowledge of the Shewhart plan-do-check-act cycle or training in the Deming method. For the Poughkeepsie manufacturing personnel involved, it was simply a natural and logical way to isolate causes of defects and prevent their recurrence.

managers readily agreed to the proposal and the concept of the ratcheting target was born. Since then, this concept was adopted at other locations and for other products within the company with excellent results.

Measurements, if performed correctly, provide a factual and quantitative basis for improvement. Referring back to Figure 5-5, the trend chart or line graph shown leads us to conclude that the documentation process is improving over time because there is a general decline in total errors. However, because errors can originate at several different steps in the process, control and improvement can only result from analyzing the process and examining the individual, or piece-part, contributions to the errors in the output.

Feedback and Corrective Action

In order to examine the individual error contributions (such as typographical errors, errors in duplication, or copying and distribution list errors), we need to establish control points at the appropriate steps in the document distribution process. At these control points, measures can be established resulting in the bar graph shown in the figure. The individual contributions of each of the three types of errors are now readily observable. At each time period, the process manager or team should provide feedback to the people involved in each of these three process steps. In so doing, the owner should be sensitive to the manner in which feedback is to be conveyed. Performed constructively, feedback is generally viewed as a positive action to improve; performed in a punitive manner (perceived or actual), measurements are viewed as a club and will be resisted.

After control parameters are selected and measurements are in place, the final step in the process control phase can be taken: feedback and corrective action.

Just as measurements serve as a basis for problem identification, corrective action is the means for problem resolution. Providing feedback and taking corrective action on untoward process deviations is of fundamental importance in both stabilizing and improving the process. Corrective action is performed with one of two objectives in mind: regulation or improvement.

Without corrective action, neither regulation nor improvement can be achieved, and the process becomes unstable and may degrade in time. Corrective action, however, cannot occur without feedback. Many business processes today are essentially unregulated; little, if any, feedback occurs. In several of the cases described in Part II, a consistent theme emerges: employees received no feedback on the quality of their work output. As a result, they were unaware that they were producing errors. Second, they did not realize the impact these errors were creating downstream in the process—a common occurrence.

Oftentimes, feedback provides a positive response. In these cases, the people realized that their outputs should have been perfect and suggested ways of foolproofing their tasks to achieve defect-free work. In other instances, such as where cycle times are concerned, a conscious decision to improve the process is made. In either event, feedback is crucial not only to regulation and control but to improvement as well.

Regulatory or control actions generally maintain a process or operation within certain prescribed limits. Improvement actions usually result in reduced variability of output quality or increased capability, or both. In either event, improvements in productivity, efficiency, and effectiveness occur, resulting in lower product or service costs and increased competitive capability. This has been seen repeatedly and is a natural outcome of improvement that is fundamental to TQM.

In summary, process management consists of three phases that comprise six basic steps:

Phase I: Initialization	Establish ownership, define boundaries and interfaces. In this phase, both process authority and scope as well as hand-offs are identified.
Phase II: Definition	Define the process. In this phase, a baseline is established from which to evaluate and assess the process.
Phase III: Control	Define control points and measurements, and control the process through feedback and corrective action. In this final phase, a means of regulation is put in place.

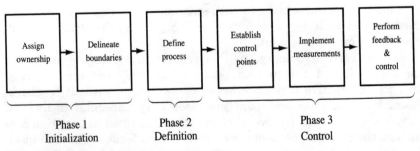

Figure 5-8.   **Phases of process management.**

These phases are depicted in Figure 5-8. Once having completed these phases, we have a foundation for analyzing and improving the process.

In the following chapters, we will examine criteria and techniques for analyzing a process. Just as process definition provides a baseline or frame of reference for improvement, process analysis assists in determining specific activities that can be improved.

Notes

1. P. Thompson, G. DeSouza, and B. Gale, "The Strategic Management of Service Quality," *Quality Progress*, June 1985.
2. J. F. Early, "Strategies for Measurement of Service Quality," 43rd Annual Quality Congress Transactions, American Society for Quality Control. Quotation reprinted by permission. ©The American Society for Quality Control, Inc. 1989.
3. C. A. Sengstock, "The Quality Issue in Public Relations," *Quality*, December, 1990. Reprinted with the permission of the Hitchcock Publishing Company.
4. *New York Times*, Jan 3, 1992, page C3.
5. J. M. Juran, Ed., *Quality Control Handbook*, First Edition, McGraw Hill, New York, 1951.
6. A. V. Feigenbaum, *Total Quality Control*, McGraw Hill, New York, 1956.
7. P. Crosby, *Quality is Free*, New American Library, New York, 1980.
8. N. F. Rockart, "Chief Executives Define their Own Data Needs," *Harvard Business Review*, March-April, 1979.
9. E. Dmytrow, "Assessing Process Capability in the Federal Government," *Quality Progress*, October, 1985.
10. W. J. McCabe, "Quality Methods Applied to the Business Process," Transactions 40th Annual Quality Congress, 1986, pp. 429–436, American Society for Quality Control.

6

Analyzing the Process: The Classical Method

We now turn to specific techniques used in process improvement. There are several approaches to examining processes for improvement, depending on whether work, materials, or information flows are involved. The most frequently used method for work flow is process analysis.

Process Analysis

Process analysis is a systematic way of defining the activities and tasks within an operation, generally at a departmental or work-group level. A classic definition of process analysis is found in an early industrial engineering handbook:

> Process analysis may be defined as the subdivision or resolution of a manufacturing process or office procedure into its constituent operations and attendant material movements so that each operation and internal handling may be studied and its necessity in furthering the process determined.[1]

Note that traditional process analysis takes into account both work activity and movement of work.

The origins of this technique trace back to the early 1900s when the Gilbreths applied Taylor's principles of scientific management to factory operations. Manufacturing processes were divided into operational elements. Each element was studied in detail together with their interrelationships. The main objective of these studies was to improve production efficiency. Cost reduction was the major benefit because most manufacturing operations involved a significant amount of direct labor and little indirect, or overhead, labor. Hence, addressing the manner in which manufacturing takes place produced immediate and tangible results. Process analysis, then, is a technique originally developed for improving product costs. Through their work in improving efficiency of production, the Gilbreths became known as efficiency experts.

Analyzing the flow of work is also fundamental in process improvement as Schroeder[2] notes,

> The study of work flow deals directly with the transformation process itself since most transformation processes can be viewed as a series of flows connecting inputs to outputs. ...When the sequence of steps used in converting inputs to outputs is analyzed, better methods or procedures can usually be found.

Process Analysis Procedure

In classical process analysis (also called process flow or methods analysis), the process is examined by going through the following steps:

1. Select the particular process or subprocess to be studied.
2. Determine the objective of the analysis—for example, reducing product cost.
3. Describe the operations involved by means of process flowcharts.
4. Assess the process and evaluate opportunities for improvement.
5. Obtain management approval for the proposed improvement.
6. Implement the improvement.

In this approach, work tasks are determined by interviewing the people actually performing them, either at their work positions or offline in private interviews. Interviewing is the main approach used when accurate and up to date process documentation does not exist. Traditional analysis involves asking a series of basic questions covering the what, who, where, when, and how of the process. Barnes[3] points out that an important factor in analysis of work is developing a questioning attitude about every aspect of work activity. Table 6-1 illustrates the various kinds of questions asked.

The next section is an example of applying classic process analysis to a service operation.[4]

Application Example

In the claims department of a large insurance company, a series of clerical activities is required to screen requests, prepare and process authorization forms to investigate claims submitted by policy holders.

TABLE 6-1   The Basic Questions of Process Analysis

1. What is the activity? What is the purpose of it? Why is it done? What would happen if it were not done? Is every part of the process necessary?
2. Who does the work? Why does this person do it? Is there somebody who could do it better? Can changes be made to permit a lesser skilled person do the work?
3. Where is the work done? Why is it done there? Could it be done elsewhere at lower cost?
4. When is the work done? Why should it be done then? Could it be done better at another time?
5. How is the activity performed? Why is it done this way? Can it be combined with another activity?

A large number of incoming claims were creating a heavy work load and a significant backlog. As a result, turnaround time had increased substantially. It was thought that this load could be reduced if the processing of these forms could be streamlined. A process analysis was performed to see if something could be done.

The analysis involved the following steps:

1. A flowchart (Figure 6-1) depicting the present process was prepared. Tasks were ascertained by interviewing employees in the department. Note that the steps shown were detailed at the task level using traditional ASME work flow symbols (described in Chapter 4). A diagram showing the physical flow of work was also prepared.

   The following steps were involved in this process: A claims clerk received a request from the Claims Department, pulled the client's file, located the claim information in the file, typed an authorization form to request a claim investigation, and inspected the completed form. The clerk then took the form to his supervisor for an approval signature and waited for the sign-off. After obtaining the signature, the clerk prepared three copies: one copy to be mailed to the regional investigator, one copy for the Claims Department (which is routed internally), and a copy to be placed in the client's file. A total of fifteen steps, including seven operations, were involved for an annual cost of nearly $400,000 for the department.

2. During a series of interviews, people familiar with the operation, including supervisors, made suggestions for better ways of processing the claims. These interviews led to the development of an improved process. The process and flow diagrams are shown in Figure 6-2.

   The simplified process reduced the number of steps from fifteen to ten: removing the Claims Department's request with the client's claim form from the in basket, typing an investigation authorization form, inspecting the three-part form, attaching a routing slip, and mailing it to the supervisor for approval. The

Task/ Activity	Description	Symbols
1	Receive and log-in request from claims department	
2	Go to filing area, retrieve client file, and return	
3	Extract claim information from file	
4	Type up investigation request form	
5	Check form for accuracy and completeness	
6	Go to supervisor's office	
7	Obtain signature	
8	Return to desk	
9	Separate copies of signed form (3)	
10	Address envelope for investigator and claims department	
11	Insert investigator's copy in envelope and put in out basket for mail pickup	
12	Insert claims department copy in envelope and put in out basket	
13	Insert third copy in client's file	
14	Go to filing area, file client's folder	
15	Return to desk	

Figure 6-1.   Process flow diagram: prepare and process authorization forms for insurance claims (present method).

supervisor then reviews and approves the claim request and returns it to the claims clerk's IN basket. The clerk then separates the three-part form, prepares the regional investigator's copy for the outside mail, puts the Claims Department's copy in the internal

Task/Activity	Description	Symbols
1	Receive and log-in request with claim form from claim department	
2	Type up investigation request form	
3	Check form for accuracy and completeness	
4	Put completed form in out basket	
5	Receive signed form and separate copies	
6	Address envelopes for investigator and claims department	
7	Insert investigator's copy in envelope and put in out basket	
8	Insert claims department copy in envelope and put in out basket	
9	Go to filing area, file client's copy in folder	
10	Return to desk	

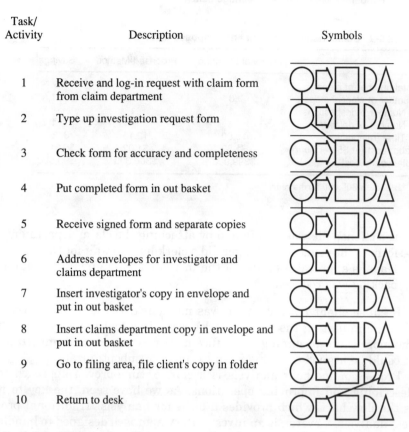

Figure 6-2.   Process flow diagram: prepare and process authorization forms for insurance claims (improved method).

mail, and, finally, puts the third form in the client's file. In this simplified process, the retrieval of the client's file at the beginning is eliminated by attaching the client's claim form to the Claims Department's request. Walking to the section leader and waiting for a signature are also eliminated through the use of interoffice mail. (Note, however, that the time for obtaining a signature may now be increased because of the transit time of interoffice mail. The claims clerk's walk-and-wait time has been eliminated and replaced by mail activity.)

3. Both present and new methods were then compared on the basis of labor cost (see Table 6-2). Improving the process resulted in cost savings of over $93,000 a year. Management approval for the change was obtained and the improvement then implemented. The productivity of the claims operation improved by reducing the time spent per claim from sixteen to twelve minutes. The retrieval of

TABLE 6-2 Comparison of Present and Proposed Methods

	Present method	Proposed Method	Savings
Total number of steps	15	10	5
Number of clerical steps	8	7	1
Number of inspections	1	1	–
Number of delays	1	–	1
Time (minutes)	16.00	12.15	3.75
Labor cost (at $5 per hour)	$1.33	$1.02	$0.31
Total yearly cost*	$399,900	$306,300	$93,600

*Based on 300,000 forms/year

the client's file was eliminated by attaching the claim department's form to the request. No longer did a clerk have to wait for the section leader's approval. Because of the increased throughput, the claims backlog was reduced.

Because client response time was not considered a critical success factor, eliminating the claims clerk's wait time and replacing it with the more time-consuming interoffice mail was not deemed important. It could be argued that this is an example of suboptimization.

Process or methods analysis is a classic, time-proven approach to describing and improving operations. As we have seen, the key to it is the flowchart, which provides a basis for analysis. Traditional process analysis, however, is an investigatory approach designed to handle simple subprocesses or sets of activities generally contained within an operation. It was intended to be used as a tool for process simplification within a department or a relatively self-contained function. It assumes that ownership is clear and unambiguous. Boundaries are well-defined and interfaces are essentially nonexistent. The analysis is generally performed by a person skilled in task analysis (such as an industrial or manufacturing engineer or a methods analyst). Improvements resulting from the analysis are usually made without regard to the social dimensions of the process. In the following chapter, other approaches to process analysis are described that answer some of the limitations of the classic method.

Notes

1. L. P. Alford and J. R. Bangs, Eds. *Production Handbook,* Ronald Press, New York, 1948. Reprinted with permission, J. Wiley & Sons.
2. R. Schroeder, *Production and Operations Management,* McGraw Hill, New York, 1981.
3. R. M. Barnes, *Motion and Time Study,* 4th ed., Wiley, New York, 1958.
4. N. Gaither, *Production and Operations Management,* 3rd ed., Dryden Press, New York, 1987.

Analyzing the Process: Modern Methods

Traditional process analysis was described in Chapter 6 in connection with analyzing a set of work activities. We now turn to methods that have been developed within the last decade that answer some of its limitations. In this chapter, three approaches to analyzing a self-contained set of activities within a work group will be described. The first is department work product analysis. Next the analysis of complex, interfunctional processes known as crossfunctional process analysis will be examined. Finally, a unique approach to analyzing service operations called the service blueprint will be reviewed.

Departmental Work Product Analysis

Departmental work product analysis is based on detailing, classifying, and examining the activities of a group. Three methods will be described in this chapter: department activity analysis, department quality analysis, and the complexity model.

Consider, first, a set of activities contained within a work group. A work group can be defined as the smallest organizational entity that is managed. This entity is commonly termed a department. A supervisor or first-line manager is generally in charge of this work group. The supervisor has the traditional management responsibilities of planning, organizing, directing, staffing, controlling, and budgeting. The degree to which each of these responsibilities is exercised is a function of the management philosophy of the firm, the style of management being practiced within the hierarchy to which the supervisor belongs, and the supervisor's own management style and disposition.

For analyzing work activities within the group, the supervisor has essentially three options. First, the analysis can be performed jointly by the supervisor and the employee. Second, the analysis can be conducted by an analyst-consultant and the employee. Third, the analysis can be done solely by either the manager or the employee. In each of these cases, the focus is on the work done by an individual employee. The traditional process analysis method described in Chapter 6 is often used in these cases.

The fourth alternative is to analyze the output of the entire work group. This alternative suggests the use of a participatory approach, which can be termed work group activity analysis.

Department activity analysis

A technique developed at IBM by P. Andreasen and A. F. Dottino called department activity analysis (DAA) is an example of work group analysis. It was developed to determine the types of activities within a department and to obtain a cost-of-quality estimate. The method was taught and used at various IBM facilities around the world. DAA has been used extensively throughout IBM to help define activities at a department level as well as identify non-value-add activities. By focusing on the waste activities that are often reflected in doing a task over, improvements in a process can be achieved in a direct and relatively quick manner.

The major tools of DAA are the input-process-output model described in Chapter 2, the customer-producer-supplier model described in Chapter 3, and cost of quality described in Chapter 5. In group meetings, each department member is asked by the supervisor to describe his or her activities. Each activity is analyzed and classified in terms of how much time is spent on "prevention," "appraisal," "failure," or "required" work. The first three terms were defined in Chapter 5 in connection with cost of quality, the last term refers to normal job activity. DAA serves as a means for defining process steps as well as performing a cost-of-quality estimate.

By examining the work product of the group, this technique:

- identifies the various output and input requirements,
- discloses how the resources in the group are being applied, and
- identifies cost and labor-saving opportunities that exist within the department and that cross departments.

In addition to these tangible advantages, an intangible benefit was noted by Parker:

> Probably the single largest benefit from DAA comes in an intangible but very evident form...the increased trust and cooperation that is developed between management and new management. DAA has contributed significantly to an increased perception on the part of new management that management really is interested in getting everyone involved in the quality improvement process.[1]

Here, Parker refers to the uncertainty on the part of management currently in position as to the intentions of higher level (new) management toward quality improvement. In a number of DAA cases that the author has been personally involved with, a notable increase in group cohesion and morale was noted—another, and perhaps more significant, benefit.

The burden of conducting the analysis is on the supervisor, so it is assumed that he or she has some flow-charting knowledge or can portray the activity in some meaningful fashion. DAA consists of six basic steps, several of which are to be conducted through department meetings with all members participating:

1. First, a list of the major activities in the department is derived. Each activity must be specific and observable. For example, in a purchasing department, expediting, if performed, would be considered an activity.

2. For each activity listed in step 1, the following are identified:

   a. The output work product
   b. The various inputs required to achieve the output
   c. The requirements of each output
   d. The requirements for each input

   Any ambiguity regarding outputs, inputs, or requirements needs to be resolved at this point with the participants. The activity description must reflect agreement on these requirements. As an example, expediting activity may include the delivery of an item based on an accelerated schedule. Inputs are the item to be expedited and accompanying information such as destination and time.

3. For each major activity, a list of the "tasks" comprising the activity is developed. For example, expediting may consist of tasks such as logging-in an expedite request, obtaining the item, contacting the individuals involved, and transporting the expedited item. (Note

that the "tasks" as defined in DAA are activities as defined in process management—see Chapter 2.)

4. The tasks enumerated in step 3 are then classified into prevention, appraisal, failure, and required work categories. The average number of hours expended in each category is then estimated.

   Tasks consistent with the mission of the work group are categorized as required work or value-add. Other tasks that are inconsistent with and do not support the mission are identified as non-value-add and evaluated with view towards elimination. Expediting, in the example used in the previous step, can be considered non-value-add, because it represents an activity inconsistent with the mission of the department.

5. Using the direct and indirect costs (labor and overhead rates) for the department, the cost of the tasks categorized in step 4 are determined. Based on the cost estimate, the key cost of quality elements namely, appraisal, failure, and prevention costs, are summarized. The elements provide a basis for evaluation and assessment. Typically, appraisal and failure costs are predominant components in a cost-of-quality estimate. Prevention costs are usually small in the initial estimate.

6. With the key elements identified, the following are determined in the final stage of a DAA analysis:

   a. Specific tasks contributing to failure costs
   b. Reasons for these tasks
   c. Means for eliminating these tasks

   For example, a purchasing department found that 30 percent of its total operating costs involved expediting certain types of components used for production. A more in-depth analysis showed that late order placement and lack of adherence to lead time requirements were the root causes. Emphasis on timely placement of orders eliminated this failure cost.

   Figure 7-1 shows an example of form sheets used by IBM for a department activity analysis. Sheet 1 is used to enumerate the activities performed in a receiving department of the Production Control function of the Charlotte plant. Sheet 2 describes the inputs, value add and output of the first activity, keypunching, listed on sheet 1. Sheet 3 describes the agreed-to input and output requirements, quality measurements used, and time spent on the activity.

   DAA, through a step-by-step procedure, provides a perspective on work activities that can be used to reduce the overall cost of quality and improve department productivity. At IBM, department activity analysis has been performed at department, function, and plant-

SHEET 1

Function Control Production Control	
Department Name Receiving, Administration, Back orders	Dept. No.

GENERAL DESCRIPTION OF WORK PERFORMED WITHIN THIS DEPARTMENT
(LISTING **MAJOR** OR ALL ACTIVITIES):

Keypunch
Transaction screening
Receiving buy/pay parts
Credit requisition activity
Daily activities report
Fill back orders
Type letters
Drop shipping
R.O.M. control
Receiving/Distribution

Manager's Signature	Date	Extension

Source: K. T. Parker, "Department Activity Analysis: Management and Employees Working Together," Proceedings of Conference of the IAQC, 1984. Reprinted by permission of the Association for Quality & Participation, Cincinnati, Ohio.

Figure 7-1.   Department activity form sheets

SHEET 2

Function Name	
Production Control	

Department Name	Dept. No.
Recieving, Administration, Back orders	

Activity:	Date:	Prepared By:
Keypunch		

INPUT

What: Unassigned Inventory: bulk fill reqs., (+) (−) delta adj. scrap transactions, new vendor P/N's & bal. changes, credit reqs. planned reqs. R.O.M., non-consumptive, header and program cards, count cards, loc. changes, grey stripes, + & − transfers, stock receipts
From: Kitting, sequence area, coordinators, receiving, finished cards, zones

VALUE ADD—WORK ACCOMPLISHED IN DEPT

Why Do: System is updated by the input of keypunched cards. It updates our inventory and vendor inventories
Value Added: The proper punches in the proper fields of each card—necessary to update system
Impact If Not Done: Loss of control: physical inventory, unassigned inventories

OUTPUT

What: Decks of transactions sorted by the header and loader cards submitted daily through screening
To: Receiving, kitting

Figure 7-1. Continued

wide levels. It has been useful in not only identifying work tasks
and activities creating waste but also in creating greater employee
involvement. DAA also represented the early beginnings of the
customer-producer-supplier concept, which was used and expanded
upon later in its "Quality Focus on the Business Process" effort.[2]

SHEET 3

Function Name Production Control	
Department Name Recieving, Administration, Back orders	Dept. No.

Activity: Keypunch	Date:	Prepared By:

What are the input requirements that you and your supplier have agreed to?   Correct cards used for various types of activities
All input fields correctly filled out (zero defects)
No missing information
All schedules for data input strictly adhered to

What are the output requirements that you and your customer have agreed to?   All transactions punched with zero errors
All schedules for data output strictly adhered to
All transactions have proper TX codes

What are the quality measurements that will show if your output meets requirements?   Tracking of ... Schedule misses
Defective keypunches
Transaction errors
Turnaround time

How many hours/week are spent on this activity? __35__ Hrs/Wk
COQ can be further classified into prevention, appraisal, and failure. What are they?

Prevention   __*__ Hrs/Wk
Appraisal   __*__ Hrs/Wk   * To be determined
Failure   __*__ Hrs/Wk
Total COQ   __*__ Hrs/Wk

Figure 7-1. Continued

Department Quality Analysis

A second approach to assessing department activity is to base group
work product analysis on management rather than employee input.
This concept is called department quality analysis (DQA) and has been

used at the General Telephone and Electronics Corporation.[3] DQA, like DAA, is also based on the input-output and customer-producer-supplier relationship models and is similar in many respects to DAA. DQA focuses on the quality-related aspects of the process, identifying elements of inputs and outputs of the transformation activities contributing to failure costs.

In contrast with DAA, a team interviews all the first- and second-line managers within the function. The interview information is then gathered and analyzed to show the input and output requirements, control points, measurements, and feedback of the various subprocesses involved. Once again, attention is given to output provided to the internal customer, inputs provided by suppliers, and transformation activities. The individual subprocess flowcharts are then assembled and consolidated into an overall department-level description. Opportunities for improvement are then identified. Fouse and Matesich[3] described this technique as follows:

> DQA is the analysis of an input-process-output model for a given activity, similar to the work simplification technique of process flow analysis. However, DQA emphasizes the process' quality-related aspects for both the input and the output sides of the model. The technique identifies task elements in the process that contribute to failure by checking to see if the agreed-upon outputs, inputs, quality measures and feedback of the performance results are present. Thus, quality improvement opportunities can be identified and prioritized....
>
> Individuals from each function in the department were appointed to the DQA team, which was responsible for completing the entire department's DQA. The team interviewed all first- and second-line managers and used the information to complete a DQA for each function. These, in turn, were compiled into a DQA for the entire department.
>
> The gathered information was then analyzed to depict the flow of requirements, measurements, feedback, and control points for the process. The results were set in the graphic format of the model and supplemented with observations regarding any weak or missing links. Finally, by linking the models of the individual processes together and consolidating the observations and recommendations, the team leader completed the overall department DQA. The published findings identified opportunities for quality improvement.

The authors point out that DQA is not a one-time application of a technique:

> DQA is not limited to a static, one-time look at an organization—it is also a strategic quality improvement tool. By repeating DQA periodically and comparing the results to previous analyses, a department can structurally and functionally evolve to continually improve the quality of its processes and to meet future needs.

Both DAA and DQA are useful for analyzing department or work-group activity. The major difference between the two is how activity information is derived, one by means of direct participation of the employees involved and the other by management perception of activities performed by the employees.

Complexity Model

A third departmental approach, known as the complexity model, provides a means for analyzing a process in terms of two types of activities: normal activities required because of the process itself and unnecessary, error-related activities resulting from the fact that an "imperfect" process exists. Unnecessary and additional work due to coping with errors is called complexity and involves both internal and external errors. Complexity reduces the time available for real work. Thus, a focus on unnecessary work will provide a basis for achieving process improvement and consequent productivity gains.

Fuller, the developer of this concept, describes complexity in the following way:

> In a typical operation or department, much of the work being done might be complexity that has been introduced by errors. Unfortunately, much of this complexity is usually not apparent to the manager of the department. We have been doing these unnecessary tasks for so long that we see them as part of the standard process.
>
> Some people have jobs that are largely the result of errors which have been introduced into the system. Consider these examples:
>
> 1. A person who opens and restocks customer returns;
> 2. A customer service representative who follows up on customer complaints;
> 3. A collector who calls customers who are late in paying for merchandise;
> 4. An expediter of late parts or products; and
> 5. An inspector who looks for defects.
>
> All people who are engaged in performing a standard process spend some portion of their time solving problems. All people make mistakes and must correct them. However, these activities are seen as normal parts of their jobs and no special notice is taken of them.... Each employee builds into this job some informal procedure to overcome the little problems faced each day.[4]

Complexity, then, is non-value-added work which reduces productivity and adds to the cost of the department work product.

In analyzing a process, Fuller distinguishes between time available for real work and time that is unavailable. Unavailable time is com-

TABLE 7-1 Examples of work components

Function/activity	Real work	Complexity
Accounting	Processing invoices	Reconciling invoice errors
Engineering	Product design	Engineering changes due to design errors
Manufacturing	Fabrication	Rework of defective product
Marketing	Promotion	Developing a strategy to overcome customer dissatisfaction

Adapted from F.T. Fuller, "Eliminating Complexity from Work: Improving Productivity by Enhancing Quality," National Productivity Review, V4N4, Autumn 1985.

posed of sanctioned time off such as breaks, work benefits, and so on. Unavailable time may be as much as 25 percent in some organizations.

Available time, on the other hand, has two basic components: time available for real work and "complexity time." As noted above, the complexity component is due to unnecessary activity and manifests itself as unproductive work that may be caused internally or generated externally. Examples of these components are shown in Table 7-1.

In a perfect process, complexity activities are nonexistent. In some cases, however, as much as 75 percent of real work-time is taken up by complexity activity; therefore, the focus of process analysis must be on unnecessary activity. Elimination or reduction of complexity, then, is the key to process improvement in this approach.

Complexity can be determined in three ways: through simply observing an activity, interviewing people involved in a process, or by means of work sampling. The first is subjective and qualitative and is dependent on the observer's objectivity and experience in assessing the overt manifestations of non-value-add activity. The second is similar to DAA except for the cost of quality determination. The third, work sampling, is a traditional tool used in job (or work) design and is based on statistical sampling. It enables an inference to be made of how much time is spent on specific work activities versus nonproductive activity based on a random sample of observations performed over a period of time. Since work sampling is generally based on attribute data, the audit sample equation in Chapter 5 may be used for this application.

Fuller describes a case example of how the work sampling technique was applied to a Hewlett-Packard sales office ordering operation.

Approximately thirty clerical and professional people worked in a Hewlett-Packard office taking orders for the company's products over the telephone. Management felt that a large amount of the work being performed was related to resolving problems caused by mistakes in processing and shipping the orders. It was decided that a study of the people's activities should take place so that management could have a better idea where the major problems were. Then, action could be taken to reduce them.

The work sampling plan was set up as follows:

1. The supervisor would wear the sampling watch.
2. When the watch beeped once every forty-two minutes, the supervisor would walk around a group of about ten people and ask each one what activity he or she was currently performing. Out of area or nonwork activities would be excluded from the study. If an employee was away from his or her desk when the supervisor came to collect data, no entry would be made for that person.
3. The study would cover a three-day period.

After three days of collecting data, the supervisor had a notebook containing 120 observations of the activities of 10 people. The date, time, and activity were recorded. Subactivities were not recorded.

The activities were then grouped by major category and counted.[4]

The data showed that 78 of the 120 observations, or 65 percent, were related to complexity, the remaining 35 percent, pertained to real work. The 120 observations were divided into seven (most frequently performed) categories of activities and a Pareto distribution determined (see Table 7-2).

Fuller assesses the findings as follows:

Three of the seven most frequent activities were judged to be part of the standard process of taking orders and therefore were classified as real work. The most frequent activity was processing merchandise that was being returned by customers. The reasons given by customers included

TABLE 7-2   Distribution of real work and complexity

Activity category	Work type	Percent of Total
1. Processing customer returns	Complexity	29.9
2. Order entry	Real work	20.9
3. Fixing order problems	Complexity	11.9
4. Changing orders	Real work	11.9
5. Expediting shipments	Complexity	10.5
6. Checking order details	Complexity	9.0
7. Taking orders	Real work	6.0

wrong product, duplicate shipment, and wrong quantity. This activity was categorized as complexity.

Upon seeing the data, the supervisor had several reactions. One was that "15 percent of the time my people are processing customer returns. This is equivalent to six people. This is far too many, and we need to first streamline the way we process returns and then see what we can do to eliminate them." Immediately, the supervisor made changes in the work procedures to improve the processing of returns. The supervisor felt that seeing the data sorted in the form of a Pareto chart helped motivate her to make the change. At the same time, a task force was formed to reduce the number of products returned.[4]

The complexity model serves as a concept for assessing the two basic components of work: real (productive) and wasted (nonproductive) activity. By measuring the proportion of real to wasted activity and analyzing the latter by source (internal or external), a quantitative basis for further action is established. An analysis is then performed with a particular focus on unnecessary activity.

Department activity analysis and the complexity model have several features in common. Both focus on the activities of a department or work group. The emphasis of both is on non-value-add or unnecessary activity. Whereas DAA arrives at costs through traditional cost-of-quality methods, complexity costs are derived directly from nonproductive labor time spent on complexity activity.

Fuller's model provides an excellent perspective on the proportion of nonproductive to productive work performed in business processes. He states: "The bulk of the work we do in most large organizations is devoted to fixing problems." By improving process quality, errors decrease and, therefore, complexity work reduces. A larger proportion of an employee's time can now be spent on "real" work with a resultant improvement in productivity.

Cross-Functional Process Analysis

For complex operations that encompass several functional boundaries of an organization, analysis is more extensive and time-consuming and requires greater coordination and planning. The process analysis technique (PAT), is a structured approach that was developed for use in larger, cross-functional processes.*

---

*PAT was developed at IBM by Jan Nordstrom of the IBM Productivity Center in Stockholm and has had extensive application in administrative and service areas in both the IBM Europe and U.S. operations. The technique was further developed by R. Harshbarger and has been taught extensively at the IBM Education Center in La Hulpe, Belgium and the IBM Center for Quality Improvement at Thornwood, New York.

Process Analysis Technique

PAT is essentially a step-by-step structured approach to defining the tasks necessary to execute a set of activities and simplifying them. Whereas DAA and DQA focus on activities within a department, PAT examines a set of interfunctional activities.[5]

Where applied, PAT has resulted in significant benefits. In one case, a twenty percent reduction in personnel and a corresponding reduction in process cycle time has resulted through the elimination of unnecessary and redundant tasks and the implementation of a simplified work flow. In other cases, cycle time improvements have approached 50 percent.

PAT requires the involvement of five types of individuals: the process owner, a consultant well-versed in process analysis, a lead coordinator (knowledgeable of the operation to be studied) who serves as support to the consultant, management associated with the process, and the people working within the process itself.

PAT, originally developed as a 13-step procedure, involves nine basic steps. The relationship among these steps and the parties involved is shown in the requirements-responsibility matrix of Figure 7-2. This type of matrix was described in Chapter 3. The procedure comprises the following steps:

Steps	PAT Consultant	Lead coordinator	Management	People
1. Decision to do analysis			R	
2. Define boundaries	A	A	R	
3. Prepare action plan and DOU	A	I	R	
4. Initiate analysis	A	I or A	R	
5. Information meeting	R	I	I	I
6. Process interviews	R	A		I
7. Analysis	R	I		
8. Present results	R	A	I	
9. Implement action	A	I	I	I

A = Assist, I = Involved, R = Responsible

Figure 7-2.   Process analysis technique requirements-responsibility matrix.
(Source: "Process Analysis Technique," R. Harshbarger, Science Research Associates-MacMillan-McGraw-Hill, Chicago, 1984. Reprinted with permission.)

1. *A management decision to have process analysis performed.* The starting point with PAT, as with other analytic approaches, is a decision to perform the analysis. However, it is an explicit requirement in PAT that this decision be made by the process owner and not a stakeholder or some other third party.

    There are several reasons for performing a PAT: a reaction to customer complaints regarding the quality of the output; recognition that the process is not operating efficiently or effectively; or response to a company-initiated improvement program. The original applications of PAT were based on reaction to a customer problem; later, its use was primarily for business process improvement.

    After the decision has been made to conduct a PAT, a process consultant is assigned to take charge of the analysis. The consultant is an individual external to the organization who is knowledgeable in methods of process analysis and facilitating group activity but is not necessarily versed in the details of the specific process to be analyzed.

2. *Define the limits of the process to be studied.* The second step is for the process owner to establish the process boundaries in order to identify the beginning and end of the process and the subprocesses to be analyzed. This step involves the identification of departments, work groups, and managers involved in the process, and a lead coordinator to support the process consultant. The lead coordinator is generally an individual who works in the area and who possesses a detailed knowledge of the process to be analyzed.

    During this step, a process team of individuals knowledgeable of the activities involved is assembled and a plan for doing the analysis formulated. Because of the extent and complexity of the process, both a strategy and a team are needed.

3. *Prepare and issue an action plan document.* After the boundaries are established and the process team assembled, a formal plan of action is documented. This is sometimes known as a document of understanding (DOU) or a memorandum of understanding (MOU). It is sometimes drafted by the consultant but is considered the responsibility of the process owner. This document should contain six items:

    a. The process to be analyzed.
    b. Objectives of the analysis.
    c. The beginning and end dates for the analysis.
    d. Identification of the various participants involved.
    e. A statement of current problems associated with the process, if any.
    f. A statement of the method and approach to be used.

After approval by the owner, the document is reviewed and signed by all of the managers involved in the process. Review and sign-off promotes involvement by management in the analysis.

4. *Initiate the analysis.* After the DOU has been completed and signed off, the consultant conducts a kick-off meeting with the involved managers. At this meeting, the consultant reviews the objectives and approach of the analysis. At this time, the managers may provide the consultant with a macro-level process description, generally in the form of flow charts, if available. In addition, the managers identify the people performing the activities in the process or "task experts" in their organization that will be interviewed by the consultant. This step was originally termed "perform a preliminary PAT." Figure 7-3 shows macro-level flow chart for an air express operation used as a starting point for developing the PAT analysis.

5. *Hold an information meeting.* Once the employees are identified, the consultant conducts an information meeting at which the process owner, lead coordinator, managers, task experts, and consultant meet to discuss the initiation of the analysis. At this meeting, the purpose and objectives of the analysis are reviewed and the lead coordinator explains what is required of the employees during the interviews.

6. *Conduct the process interviews.* The consultant then interviews the task experts and develops a process flow based on their description of the tasks performed. Interviews are usually done on an individual basis. During each interview, the following types of questions are asked by the consultant:

- What do you do when you perform the task?
- How long does it take?
- What support is needed?
- How often is the task done?
- What can be done to improve the task?

A lead coordinator usually attends the interviews to provide technical support to the consultant and provide any background information or explanation that may be required. At the end of the interview, the consultant has a flow diagram of the set of tasks performed by the employee; the consultant reviews the diagram for correctness with the person interviewed and finalizes it. After all the interviews are complete, the consultant is ready for the analysis phase.

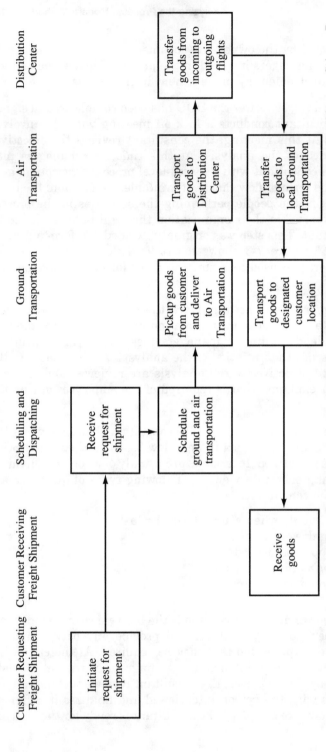

Figure 7-3. Macro-level PAT flow diagram for an air express company. (From "Process Analysis Technique," R. Harshbarger, Science Research Associates, MacMillan-McGraw-Hill, Chicago, IL, 1988. Reprinted with permission.)

Step 1. Define the process as currently performed in terms of tasks.

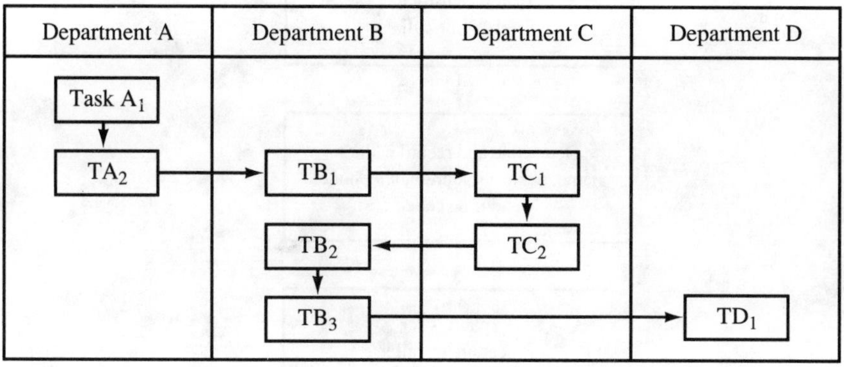
Department A
Department B
Department C
Department D
Task A1
TA2
TB1
TC1
TB2
TC2
TB3
TD1

Step 2. Perform an analysis of the existing process.

Analysis: Question all parts of the process.

- Added value of the task?
- Is the task needed?
- Could another area perform the task?

Step 3. Based on the findings of the analysis, modify the existing process.

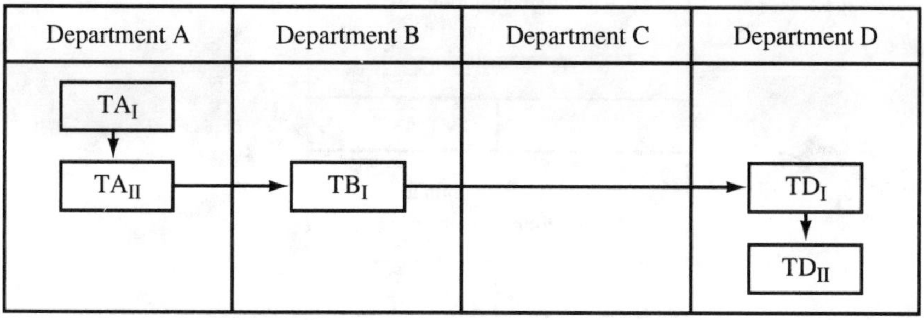
Department A
Department B
Department C
Department D
TAI
TAII
TBI
TDI
TDII

Figure 7–4. The IBM process analysis procedure.

Figure 7-4 shows the basic PAT procedure and function flowchart format and Figure 7-5 shows a task-level flowchart used for the air express operation depicted in Figure 7-3. Here the data-entry set of activities is expanded in greater detail for the scheduling and dispatching subprocess. Task times are noted.

7. *Analyze and revise the process.* Upon completion of the interviews, the consultant has developed a detailed picture of the process. The consultant and lead coordinator analyze the assembled set of tasks for improvement. As the analysis develops, the consultant questions all parts of the process and considers alternatives.

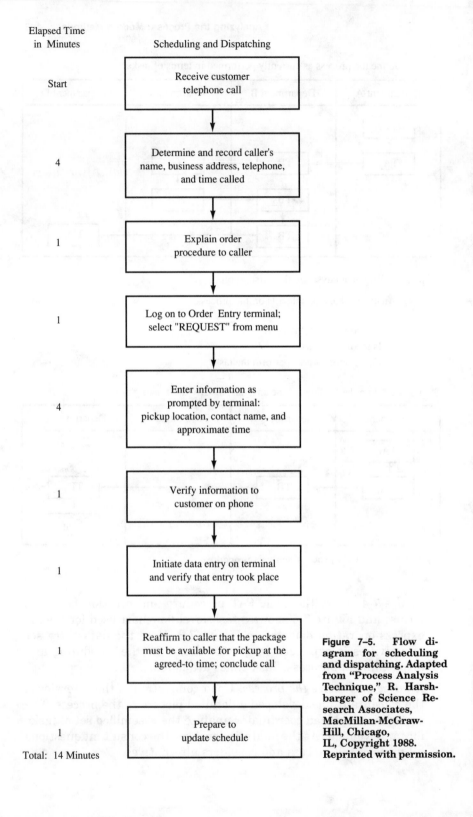

Elapsed Time in Minutes	Scheduling and Dispatching
Start	Receive customer telephone call
4	Determine and record caller's name, business address, telephone, and time called
1	Explain order procedure to caller
1	Log on to Order Entry terminal; select "REQUEST" from menu
4	Enter information as prompted by terminal: pickup location, contact name, and approximate time
1	Verify information to customer on phone
1	Initiate data entry on terminal and verify that entry took place
1	Reaffirm to caller that the package must be available for pickup at the agreed-to time; conclude call
1	Prepare to update schedule
Total: 14 Minutes	

Figure 7–5. Flow diagram for scheduling and dispatching. Adapted from "Process Analysis Technique," R. Harshbarger of Science Research Associates, MacMillan-McGraw-Hill, Chicago, IL, Copyright 1988. Reprinted with permission.

Working with the lead coordinator, the consultant distinguishes essential from nonessential tasks, tasks that can be eliminated or combined with others, and ones that can be sequenced in a more efficient fashion. When this analysis is complete, a revised process flow diagram is prepared by the consultant and recommendations for improvement developed.

8. *Present the revised process for management approval.* After the revised process is ready, the consultant and lead coordinator meet with all the participants: the process team, the owner, managers involved, and the employees interviewed to review the results of the analysis and proposed recommendations. The original and improved process flows are presented, highlighting the revisions to the original process and the benefits of the latter. As a result of this meeting, either revisions may be required or a decision by the owner to implement the recommendations may occur.

9. *Establish an implementation strategy.* Upon approval of the revised process, the process team establishes an implementation strategy that is presented to the owner for approval. The process then proceeds into an improvement phase.

These nine steps are described in detail by Harshbarger (see note 5). As can be seen, PAT is not a straightforward, traditional process analysis. By virtue of the various communication steps taken, the involvement of management, and the employment of a lead coordinator, certain social, behavioral, and organizational elements embedded in a process are recognized and dealt with.

The entire analysis can be done in six to eight weeks of part-time work for a relatively complex process. Although PAT has been used largely at IBM in connection with its quality improvement activity, other companies have used it successfully as well.[6]

Service Blueprint

A somewhat different approach to process analysis is a technique developed by G. Lynn Shostack called service blueprint. The service blueprint technique is designed primarily for service and administrative functions. Service blueprint is a four-step procedure that involves:

1. Process identification

2. Isolating fail points

3. Establishing a time frame

4. A profitability analysis

These four steps, according to Shostack, give managers a framework for looking at all the factors involved in either managing or creating a service. There are three unique features that differentiate it from other analysis techniques: distinguishing process steps that are transparent to the customer from those that analysis techniques: distinguishing process steps are invisible, designating points of possible failure, and performing a profitability analysis of the process.

The service blueprint procedure begins with laying out, schematically, the various steps or activities that constitute the particular service to be analyzed. The process map is essentially a work-flow diagram that is divided into two areas: process steps that are visible to the customer and those that are not. A demarcation line called "line of visibility" separates the two areas. Figure 7-6 is a service blueprint process diagram for a dry cleaning service. The reason for distinguishing visible from invisible process steps is that changing the latter may affect the way customers perceive the service being performed. In dry cleaning, for example, the failure to remove spots could affect a customer's perception of service quality.

The level of detail at which the process is defined is left to the person analyzing the service. Shostack states:

> Even within the simplest process, further definition is beneficial; in shoeshining, it might be useful to specify how the proprietor will perform the step called "buff"...identifying the components of a step or action reveals the inputs needed and steps covered and permits analysis, control and improvement.[7]

The second step involves determining actual or potential fail points. Fail points are steps in the process where errors in service are or can be created. In the example illustrated in Figure 7-6, inadequate cleaning is a potential fail point. By noting the various points of failure, the analyst has taken into account all of the possible steps that can affect the quality of the service. Management can then select the most critical fail points for actions to assure that failures at these points do not occur or are minimized. Identification of points of failure and design of fail-safe processes are unique in this approach to service process quality.

Below the line of visibility, there are two major fail points in this example. Because the explicit service involved is providing properly cleaned and pressed garments, the clean and press operations are critical to providing this service. As long as these operations are not, in a practical sense, fool-proofed (by, for example, automated techniques), these fail points will continue to exist. Failure to provide properly cleaned garments may be minimized by (1) warning the customer that certain types of spots may not be removable, and (2) checking the

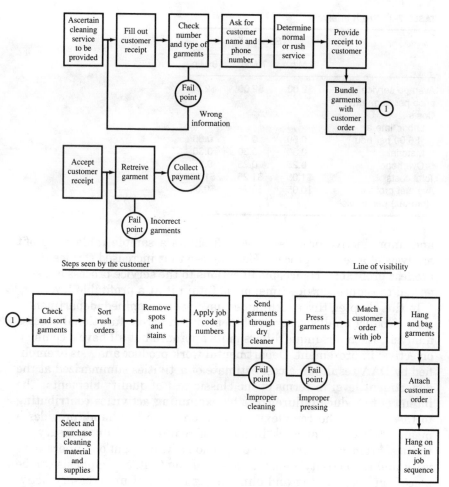

Figure 7-6.   Service blueprint for a drycleaning operation.

garments both before and after the cleaning operation. Where a troublesome stain has not been removed, failure to do so should be noted on the store ticket and communicated to the customer during order pick-up.

The third step is determining the total process or execution time of the service. A standard execution time for each activity is established first, then an acceptable latitude of time in performing the service is determined. These time values serve as a basis for determining process costs that affect the profitability of the service operation.

Finally, a table showing the revenue-cost calculation as a function of execution time is filled out and profit is analyzed. From this, an execution time standard is established to help preclude an unprofitable

TABLE 7-3   Profit analysis

	Service time		
	5 Min.	7 Min.	8 Min.
Average service price per garment	$2.00	$2.00	$2.00
Costs			
Labor time at $6.00 per hour	0.50	0.70	0.80
Material	0.30	0.30	0.30
Overhead	0.25	0.25	0.25
Total costs	$1.05	$1.25	$1.35
Avg. net profit (pre-tax) per service	$0.95	$0.75	$0.65

and unproductive process. Table 7-3 shows a sample table for profit analysis of a service process. Here we see that increased service times are less profitable. Hence, modifications to the service process are suggested to reduce service time, or maintain it at a profitable level.

In this chapter, three techniques have been examined: departmental work product analysis, service blueprint, and cross-functional process analysis. Each technique differs in its approach but all have a common objective: improvement. Departmental work product analysis exemplified by DAA results in a cost estimate of activities summarized at the department level in terms of the classic cost of quality elements. The intent is to reduce failure costs by examining activities contributing to these costs. The complexity model, on the other hand, provides a means for looking at work in terms of required and unnecessary activities. From determining the proportion of time spent on unnecessary work and the activity elements contributing to it, an emphasis can be placed on determining and eliminating causes of unnecessary activity. Shostack's blueprint approach, in contrast, provides an estimate of profitability of an operation and allows a focus on potential fail points in the service process. Finally, process analysis technique is a structured approach designed for complex, cross-functional processes. The emphasis is on improvement by reducing or simplifying the various activities contained within the process.

Notes

1. K. T. Parker, "Department Activity Analysis: Management and Employees Working Together," Proceedings, 1984 IAQC Annual Conference. Reprinted with the permission of the Association for Quality and Participation, Cincinnati, Ohio.
2. E. S. Kane, "IBM's Quality Focus on the Business Process," Quality Progress, April, 1986.

3. F. E. Fouse and J. A. Matesich, "Department Quality Analysis: A Case Study," Quality Progress, June 1987.

4. F. T. Fuller, "Eliminating Complexity from Work: Improving Productivity by Enhancing Quality," National Productivity Review, V4N4, Autumn 1985. Reprinted with the permission of Executive Enterprises, Inc., 22 West 21 St., New York, NY, 10010. All rights reserved. Copyright 1985.

5. R. Harshbarger, "Process Analysis Technique," Science Research Associates, Macmillan-McGraw Hill, Chicago, 1988.

6. J. J. Entwistle, "Business Process Management at Ortho Diagnostic Systems, Inc." Proceedings IMPRO 89, Juran Institute, Wilton, Connecticut.

7. G. L. Shostack, "Designing Services that Deliver," Harvard Business Review, Jan/Feb 1984.

8

Assessing and
Evaluating a Process

In the previous chapters the fundamentals of process management were described. We saw that a process can be examined in terms of three phases: initialization, definition, and control. But once a process is in place and operating, how does one know how well it is functioning? In this chapter, we will look at ways of assessing and evaluating a process.

Assessment is performed to determine the state of a business operation. Much as operational audits determine compliance to procedures, assessment determines how well the process is being executed. Process assessment involves establishing criteria for evaluation, performing the evaluation, determining strengths and deficiencies of the process, and drawing conclusions regarding its state. Evaluation criteria provide standards for judging a process. Assessments can be performed singly or by teams of people; they can be performed by people currently working within the process or by people external to it. Evaluations are generally performed according to some preplanned agenda or checklist.

Any business process has certain identifiable strong points and weaknesses. The result of an evaluation is a better understanding of its strengths and deficiencies. Consider, for example, a marketing process having pricing as a subprocess. An assessment may show the cost-estimating part of this subprocess to be quite accurate. Accuracy is, therefore, noted as a strong point. However, the speed by which a cost estimate is generated is slow in terms of market requirements. The assessment would indicate that cost estimate response time is a deficiency or weakness. Strengths and weaknesses can be used to characterize the basic attributes of a process: effectiveness and efficiency.

These two attributes serve as the principal criteria for evaluating a process.

Criteria for Assessing a Process

There are two prime criteria for process assessment: effectiveness and efficiency.* In process management, effectiveness refers to how well the output meets customer requirements—a measure of actual against intended output. Efficiency is a measure of how well the internal operations are performed in terms of output and resources required to achieve the output. Process assessment must take into account both effectiveness and efficiency.

Effectiveness and Efficiency

In managing a process, one must first assess whether or not it is effective. Once having achieved effectiveness, attention is next given to making the process more efficient. In essence, efficiency follows effectiveness. By improving the process to make it more effective, it may simultaneously become more efficient. For example, in modifying a process to improve its cycle or response time, several repetitive tasks that contribute to the overall process timing may be eliminated. By eliminating them, the resources applied to these tasks are also eliminated. Both efficiency and productivity improvement result from this.

The following symptoms are indicative of an ineffective process:

- Customer complaints
- Inconsistent output quality
- Lack of awareness of output quality
- Absence of corrective action
- Lack of interest in the customer
- Long response times in reacting to a problem

Process effectiveness is best assessed by measurement, both internal and external. External or output-based measurements must re-

---

*A third criterion, adaptability, is also used. Adaptability refers to the ability of a process to adjust to change—either technological or output changes. A process that requires significant changes in either capital equipment, personnel, building, or all three is not considered adaptable. For example, suppose management wishes to change the personnel of a billing process from individual operations to group operations using empowered teams. The employees by nature are individualistic and cannot work as members of a group even after training. This indicates poor process adaptability.

flect customer requirements; internal measurements must reflect an adequate translation of both external and internal customer requirements. This translation should proceed from the external customer boundary all the way to the starting activities creating the product or service. Deployment matrices (described in Chapter 3) in connection with establishing requirements are a useful method for performing this translation.

Efficiency, on the other hand, reflects productivity of the internal operations or how well resources are utilized in the process. A traditional measure of efficiency is the ratio of actual output to the effective (i.e., practical) capacity of an operation. Suppose, for example, that the effective capacity of a document processing operation is 1,000 documents a day. Output is measured at 800 a day. Efficiency, then, is 0.8 or 80%.

Improving a process often improves its effective capacity and, hence, its output capability. Conversely, process improvement may create increased output with less resources. This implies improved productivity. In producing an output, productivity and efficiency are directly related. Productivity, P, is a ratio of output to input and is measured through the following equation:

$$P = \frac{O}{I}$$

where I is the input and O is the output of a specific operation or set of operations. Productivity can be measured either in terms of labor or economic value. For labor, productivity is expressed in terms of output per unit of labor such as number of cars produced per man-hour.

The second type of productivity measure is known as multifactor productivity, which is used when it is desirable to reflect other resources besides labor. Here, costs are often used so that a multifactor productivity becomes, for example:

$$P = \frac{\text{Output value}}{\text{Input costs}}$$

$$= \frac{\text{Price} \times \text{output quantity}}{\text{Labor cost} + \text{materials cost} + \text{overhead cost}}$$

Clearly, efficiency and effectiveness are not always directly correlated. An effective process may be very inefficient and an efficient process may be ineffective. For example, a customer requirement may be defect-free documentation. However, in order to achieve this, the process owner has put in place three successive document inspections done to assure this requirement. This is an effective but not very efficient process: the inspections may eliminate document defects but the added resources required to accomplish this reduce the efficiency of the operation.

To illustrate an inefficient and ineffective process, consider the following. A is a supplier to B. A may be a manufacturer of hardware or a supplier of information of some kind. B is required to transform the input into an end item of higher added value. As a consequence of a previous problem, B has reacted and provided an inspection step on the input side of B's process—a common action. There is little or no communication between B and A on the quality of A's output.

The result of installing an inspection in B's process is the creation of both an ineffective and an inefficient process. It is ineffective in that A's output has not met B's requirements. Moreover, it is inefficient in that B has had to provide additional resources to inspect A's output in order to meet B's requirements, resulting in reduced productivity in B's operation. Because of the lack of communication between A and B— often because of the effects of organizational boundaries—A may not be aware of the quality of its output. Hence, the lack of feedback and subsequent defensive action in response to a problem has become the root cause of inefficiency and ineffectiveness—a common phenomenon.

Some of the symptoms of an inefficient process are:

- The existence of numerous verification or inspection operations
- Redundant, unnecessary, or non-value-add activities
- Numerous corrective activities such as rework and reconciliation
- Chronic input or supplier problems
- Excessive costs of value-add activities

In the case of the first symptom, technology constraints may dictate the need for numerous verification or inspection operations to achieve a required level of defect-free output. This is especially true in complex processes such as semiconductor fabrication and software development, where numerous inspections are performed. On the other hand, the existence of many inspection steps in an administrative process may reflect the institutionalization of control points without regard to process efficiency or cost.

Although it is rather obvious that a redundant activity is inefficient, it is not always apparent that redundancy exists in an operation. Redundancy becomes obvious when the process is defined. For example, an order or request may be first inspected at department X and again at department Y later in the process. This type of redundancy is seldom evident until a process flowchart is developed.

Rework of defective material or the correction of errors in information is also symptomatic of inefficiency. These are non-value-add activities that cause waste of resources. Chronic input or supplier problems range from late and incomplete shipments of material or transmittal of information to the wrong item being sent.

Excessive costs of value-add activities are costs due to duplicate certain conditions, such as protective inventories for an improbable event, and recycling due to trial and error activity (i.e., not getting it right the first time). An efficient process produces the required output at minimum cost. On the other hand, an inefficient process produces the output at a higher cost.

A cost-of-quality evaluation (described in Chapter 5) can provide an assessment of process efficiency. This can be done by examining individual activity contributions to failure and appraisal costs. For example, substantial failure costs due to scrap or rework and repair can be an indicator of inefficiency.

Conti[1] has proposed a process performance measure that is a ratio of the cost of activities that create added value from a customer's viewpoint (CVA) to the total process cost (Cp). Efficient processes have high CVA/Cp ratios (values approaching one) and, conversely, inefficient ones have low ratios. Thus processes containing a large number of non-value-add activities such as rework, inspection, and expediting generally reflect process inefficiency.

Process Evaluation: Rating Method

A useful approach for evaluating a process is the rating method developed by IBM in connection with its business process improvement efforts.[2] The rating method is based on five levels or stages of process management and proceeds from a rating of 5 to a 1 in terms of increasing maturity:

Level 5 indicates that there is no designated owner, process management is nonexistent, and the process itself is ineffective. In addition, major deficiencies may exist that require corrective action.

Level 4 signifies that the basics of process management are in place, certain improvements have been identified, and a corrective action plan established. To achieve this level, the following eight items must be accomplished:

- A process owner must be identified and designated.

- Customer-supplier relationships and requirements (both internal and external to the process) must be established.

- The process must be defined and documented.

- Control points within the process must be determined.

- Measurements of effectiveness and efficiency must be identified and put in place.
- The process must be assessed and deficiencies or exposures (such as defects, rework, excess cost, redundancies, supplier problems, etc.) identified.
- Statistical methods must be implemented and data collection is under way.
- A defect prevention methodology must be in place and a feedback mechanism established for continuous quality improvement.

Level 3 means that the following elements have been accomplished in addition to the ones listed above:

- Process effectiveness measurements show evidence of meeting customer requirements.
- No significant control exposures exist.
- Improvement items to achieve level 2 have been identified and a plan set in place to achieve this level.

A business process at this level is effective. Some operational inefficiencies or control exposures may exist but are not considered critical. At IBM, many business processes achieved this level.

Level 2 signifies that major improvements in the process have occurred and positive results have been realized in terms of both effectiveness and efficiency. It is also adaptable to future demands placed on the operation. To achieve level 2, all elements comprising levels 3 and 4 must be in place in addition to the following:

- Efficiency measurements that demonstrate continuing improvement in terms of reduced resources per unit of output.
- The process is assessed as competitive both in terms of effectiveness and efficiency as compared with comparable processes within the firm or industry. In other words, the process is benchmarked (benchmarking is discussed later in this chapter).
- The process is adaptable to business direction changes without loss of efficiency and is deemed by the customer to be able to meet requirements for several years. (At IBM, this is taken to mean the time duration or period of the strategic plan.)

Level 1 business processes operate at maximum efficiency, are benchmarks or leaders in this type of operation, and function at maximum effectiveness. To achieve this top rating, the requirements for achieving 4, 3, and 2 ratings must have been fulfilled in addition to the following:

- The output has been assessed to be primarily defect-free.
- The process operates with minimum resources.
- The process is considered "best of breed," that is, one that can be used as a model for benchmarking.

Few processes at IBM achieved this level but those that did achieved outstanding results. An engineering design process at IBM Kingston, for example, reported a defect-free operation for more than two successive years.[3] This process is described in Case 3, Chapter 13.

A questionnaire useful for evaluating a business process in terms of maturity levels is provided in the appendix to this chapter.

Example

An example of a process evaluation employing this rating method is the following report on a financial subprocess involving invoice reconciliation in a multinational financial operation of a major firm. The assessment was performed by the management of the function in response to a request from corporate headquarters. The owner's name is disguised.

Process evaluation

Process name: Invoice reconciliation—international operations

Owner:	Q. W. Smith
Department:	International billing
Date:	4/15/92

Customers	Suppliers
U.S. headquarters finance department	European countries finance departments
European headquarters finance department	U.S. billing departments
European countries: Finance depts. and plants (U.K., France, Germany)	European headquarters finance department
	U.S. headquarters finance department

Control Measures
- Invoice aging statistics by country and invoice type
- Reconciliations
  - Dates received and returned
  - Overdue statistics
- Monthly control indicators
  - Number of unreconciled invoices (current accounts)
  - Number of invoices over 270 days

Process description. The reconciliation process receives inputs from two activities. The first is invoices that result from the daily shipment of goods to the European plants. These invoices originate from customer brokers performing this service and are sent to international billing as well as to the European country receiving the goods. The country enters the verified invoice onto its payables account.

The second input is a monthly billing from the U.S. headquarters finance department, a copy of which is also sent to the European headquarters finance department. The European headquarters finance department, in turn, bills the receiving country for goods received for the month. The country enters these charges against their original entries, which reflect the copy of the broker's invoice.

A list of entries that do not match or agree is a discrepancy report known as a *current account*. A discrepancy report is generated monthly by country and is sent to the international billing department for reconciliation. This department investigates each discrepancy and responds to each country with its findings annotated on the discrepancy report.

An updated desk procedure is in place to ensure that a uniform procedure is followed by each accountant involved in reconciliation.

Process concerns.

- Number of unreconciled invoices (current accounts)
- Number of invoices exceeding 270 days

Assessment: level 3. The process as currently practiced is effective, and no significant operational deficiencies exist. Defect-free criteria have been established and quality improvement activity is in process. The improvement activity under way will ensure the achievement of defect-free status in the department's reconciliation process.*

Action plans.

1. Utilize electronic mail. This will improve the transit and response times of the process. Checkpoint: 10/1/86.

---

*The reader, after reviewing the criteria for each level, would probably question management's assessment that this is a level 3 process. It is more likely a level 4 since there is no evidence that their measurement criteria are customer-based. Management perceived that they were based on customer requirements but, in fact, they were traditional financial measures that had been in place for some time.

2. Conduct an intercompany billing workshop for all participants in the process. Target: 2Q87. This will improve overall understanding of the billing process among the participants and highlight points of concern and issues requiring resolution.

Adaptability

The rating system described above addresses both process effectiveness and efficiency and an additional characteristic: adaptability. Adaptability is deemed important to IBM as a process characteristic because of the effect of changing technology on business processes and on human resources. Processes have varying degrees of adaptability. Some are highly flexible—they can readily be modified to utilize a new technology and the personnel involved can switch to the new process with relative ease and minimum impact. Other processes appear inflexible; changes occur with difficulty, if at all, and require significant investment in resources to modify them to meet new requirements.

Adaptability encompasses the response of a process to changing conditions such as output requirements, internal constraints, and input quality. A process is adaptable if activities can change within the operation to meet new requirements without significant modifications. Adaptation may involve some activity, personnel, and equipment changes, but the process remains largely intact. Processes that lack adaptability are limited in some manner. These limitations may involve equipment capability, capacity, throughput, and cycle times and human aspects such as skills, flexibility, resistance to change, and other behavioral factors.

Other Evaluation Methods

Effectiveness, efficiency, and adaptability are useful general criteria for assessing a process. However, these criteria do not address internal characteristics of the operation itself. These can be assessed through one or more of the following methods:

- Performance evaluation
- Capability studies
- Benchmarking
- Quality profile

Performance evaluation

Performance evaluation addresses variability of the process. Evaluation can range from a quick check to a long-term analysis of operational performance. A simple statistical analysis may show the degree of variability that exists for a specific measure and how well the process is centered over a period of time. Frequency distributions, Pareto charts, scatter diagrams, trend charts and, in some cases, control charts may be used. The following factors can affect operational performance:

- Materials quality and flow
- The basic design of the product or service
- The operations or processing steps employed
- The quality, availability, and design of equipment and tools used
- Technical criteria and control information used
- Skill level, attitude, and capability of people employed
- Management style, attitude, and philosophy

Both output quality and process stability can be affected by any or all of these elements.

Process capability

Process capability involves a determination of whether or not a process can achieve certain operational criteria. It may also involve a determination of its operational limits. For example, measures of average, minimum, and maximum response time of a process taken over a period of time reflect its operational capability in terms of its cycle time.

Traditional process capability studies involve a statistical assessment of parameter data gathered under operating conditions and compared with specifications. These studies serve as a basis for improvement as well as an indicator of variability. The reader may refer to texts on quality control or statistical process control for further discussion of process capability studies.

Benchmarking

The concept of benchmarking provides a useful means for assessment as well as for establishing standards for process excellence. It is essentially a formalized search for the best practices that contribute to superior company performance in disciplines such as marketing, manufacturing, and distribution. Benchmarking, a concept introduced by Xerox, is a structured approach that involves investigating industry best practices, analyzing and evaluating one's own operation for opportunities, and implementing an action plan that includes revision of goals, objectives, and operating targets. Benchmarking assumes that known standards exist and can be used for comparison.

Benchmarking is divided into two parts: practices and metrics. Practices involves setting up an organization to understand and analyze the methods that are used in the process selected for benchmarking. For example, to evaluate its warehouse operation, Xerox analyzed L. L. Bean's as a standard of comparison.[4] Prior to Xerox's decision, Bean, in turn, had performed a benchmark analysis of its operation in comparison with that of several other firms. As a result, Bean modified its process, adapting portions ("best practices") of it from its knowledge of other firms. Practices such as electronic ordering between a store and distribution center, bar code labeling, and scanning were adopted. Table 8-1 illustrates, for part of the order entry process, the best practices for each of the features required.

The second part of the benchmarking process is metrics. Metrics is the quantification of the practices that have been adopted and involves setting specific criteria or targets and measuring performance. For warehousing, measures such as orders per person per day, shipments per day, and capacity utilization (square foot per stocking unit) were used. The most popular measures used in benchmarking are costs, response or cycle times, throughput, and utilization.

TABLE 8-1 Benchmarking: Best Practices Illustration

Feature	Best practice
Customer contact	■ Toll-free number ■ 24-hour availability ■ Mail ■ FAX
Requested transaction	■ Special quantity discounts—immediate quotes ■ On-line stock availibility information ■ Inquiries ■ Orders ■ Sales tax, freight information
Order recording	■ Direct update of data bases ■ Order queuing
Credit approval	■ Electronic credit card authorization ■ Line-of-credit feature ■ Check clearing and approval
Transaction confirmation	■ Inventory reservation ■ Pass/fail credit and validation

Note: Adapted from R.C. Camp, *Benchmarking,* ASQC Quality Press, Milwaukee, 1989.

Both interest in and use of benchmarking in recognizing and adopting best-of-breed processes have increased significantly in recent years. Many companies have used benchmarking to improve the performance of their business processes. Among them are, in addition to Xerox, ALCOA, AT&T, IBM, and Milliken. AT&T has described it as a "structured discipline for analyzing a process to find improvement opportunities."[5] A contributing factor to this increase has been the incorporation of this concept in the Baldrige Award criteria.

It should be noted that benchmarking, as an informal method, has existed among and within organizations for many years. Professional societies at regional, national, and international meetings exchange information both formally and informally on new concepts, methods, and techniques as well as technological innovations. Information exchange is an informal way to become aware of best practices. It does not have the structure of benchmarking, however.

Within organizations, similar information exchange mechanisms exist. It is particularly useful in organizations that are decentralized. The nature of the technology or technologies that a firm deals with is another factor. High technology organizations where technology is progressing rapidly (such as in the semiconductor industry) require frequent exchanges simply to avoid becoming obsolete.

Quality profile

Another technique for process assessment is the quality profile. The quality profile is based on the factor analysis technique employed in operations management. The profile can be used to assess the overall quality of a product or service, process, function, or even a firm. It provides an index or numerical value based on an aggregate score of the key parameters describing the attributes of the product or service, process, or organization of interest. This technique is based on the following procedure:

1. First, determine the main attributes or critical success factors governing the item being assessed. Reliability, ease of use, timeliness of delivery, and effectiveness are examples of product and process attributes that may be used.

2. Next, after all of the key factors have been determined, a weight is assigned to each factor. The factor of greatest importance will receive the greatest weight, the second highest factor, the next highest weight, and so on. The sum of the weights must equal 1.0.

3. A rating scale is then established. Scales of 0 to 1, 1 to 10, and 1 to 100 are common in factor analysis. Each factor is then ascribed a score or rating value by the assessor based on the scale used.

4. Finally, each score is multiplied by the weight corresponding to the attribute. The result of each multiplication is noted in a separate column for weighted scores. The weighted scores are then added to get a total score which now becomes a numerical value for the quality profile at a point in time.

Table 8-2 shows the layout of a generic profile and how the score is calculated. A score, then, is a value based on the factors ascribed, the weight distribution assigned, and the rating values assigned. Doing this periodically will yield a quality profile trend. Table 8-3 shows an improvement trend in a process over a period of two quarters. Yield, cycle time, throughput, and output quality were chosen as critical success factors with the weights assigned as shown.

Profiles of this type are inherently subjective because of the scoring aspect of the method. Assigning a score is basically judgmental and, therefore, prone to bias. If scoring is done by one person, the same individual should be used for each period to eliminate differences in bias between individuals. The preferred method is group scoring, in which a score is achieved either by consensus or by taking an average of individual profile parameter values determined by each person of the assessment group.

Process assessment serves both as a diagnostic and auditing tool for determining the health and maturity level (in terms of progress) of an operation. It is also useful in serving as a basis for improvement. In some cases, a simple performance evaluation may suffice for an assessment. In others, capability studies are desirable from a comparative point of view, or benchmarking may be appropriate. In tracking

TABLE 8-2   Generic quality profile

Profile parameter	Weight	Score	Weight × score
P1	W1	S1	W1 × S1
P2	W2	S2	W2 × S2
P3	W3	S3	W3 × S3
⋮	⋮	⋮	⋮
Pn	Wn	Sn	Wn × Sn
		Sum =	

Notes:
(1) The sum of the weights must equal 1.0.
(2) The same scoring scale must be used over several time periods, e.g., 0-1, 1 to 10, 1 to 100, etc.
(3) The quality profile is given by the sum of the individual weighted scores: $W1 \times S1 + W2 \times S2 + W3 \times S3 + \ldots + Wn \times Sn$.

TABLE 8-3 Process quality profile

Process parameter	Weight	First quarter Score*	First quarter Wtd Score	Second quarter Score*	Second quarter Wtd Score
Yield	0.3	6	1.8	7	2.1
Output quality	0.4	5	2.0	7	2.8
Throughput	0.2	3	0.6	5	1.0
Cycle time	0.1	2	0.2	4	0.4
		Sum =	4.6		6.3

*Scale: 1 - 10.

the critical performance attributes of a process with time, the profile method may be justified.

The preceding chapters have served to develop the foundations of process management in terms of three phases: process initialization, definition, and control. This chapter has portrayed ways of assessing a process that, in turn, forms a basis for evaluating improvement. In the next two chapters, we will look at how process management is practiced.

Notes

1. T. Conti, "Process Management and Quality Function Deployment," *Quality Progress,* December, 1989.
2. E. S. Kane, "IBM's Quality Focus on the Business Process," *Quality Progress,* April 1986.
3. E. H. Melan, "Focus on the Process: Key to Quality Improvement," Proceedings 42nd Annual Quality Congress, May, 1988.
4. R. C. Camp, *Benchmarking,* ASQC Quality Press, Milwaukee, 1989.
5. K. Bemowski, "The Benchmarking Bandwagon," *Quality Progress,* January, 1991.

Appendix: Process assessment questionnaire

The following questionnaire may be used by an assessor or an assessment team to determine a process operating at level 4 when all questions in the following section can be answered yes except for the last question, where either yes or no is acceptable for this level.

		Yes	No
1.	Does it have an owner who is fully accountable for the process?		
2.	Have the process boundaries been defined in written form?		
3.	Have the suppliers of the process and their outputs been specified?		
4.	Have the customers of the process and the outputs provided to them been specified?		
5.	Has the process been defined and documented?		
6.	Are process measurements adequate to determine whether or not the customers' requirements are being met?		
7.	Are adequate controls in place?		
8.	Do major process deficiencies exist?		

The process can be considered effective and operating at level 3 when all questions pertaining to level 4 are answered satisfactorily and all of the following questions can be answered yes.

		Yes	No
1.	Have agreed-to, defect-free criteria been established?		
2.	Have measurement targets been adjusted to reflect defect-free criteria?		
3.	Has the owner initiated process improvement actions aimed at achieving defect-free targets?		
4.	Are all defined process measurements being performed on an ongoing basis?		
5.	Is the process owner using measurements as feedback for ongoing inprovement?		
6.	Does a plan to achieve level 2 exist?		
7.	Do any significant control exposures exist as confirmed by measurements or independent reviews?		

The process can be termed as efficient and adaptable and operating at a 2 rating when all of the questions pertaining to levels 4 and 3 can be answered satisfactorily and all questions in the following section can be answered yes.

		Yes	No
1.	Have the customers specified their future requirements of the process?		
2.	Does the owner have a mechanism for identifying the impact of new business directions on the process in question?		
3.	Does the owner have an understanding of how the efficiency, productivity, and resource requirements of this process compare with similar processes?		
4.	Have the owner and the process customers come to an agreement as to what the process must deliver and how it must be adapted and changed?		
5.	Are the owner's resource constraints identified and being addressed as needed?		
6.	Are process improvements being applied to refine the process flow, procedures, activities, measurements, and control points consistent with future requirements?		
7.	Have major improvements been made to the process with tangible and measurable results realized?		

The process will be defined as defect-free and operating at a 1 rating when all questions pertaining to the previous levels are answered satisfactorily and the following questions may be answered yes:

		Yes	No
1.	Is the process considered superior to other similar processes?		
2.	Has the process undergone an independent review with a satisfactory rating in all areas and did the reviewer assess the process to be substantially defect-free from the customer's viewpoint?		
3.	Does the process operate with minimum resources for its rated capacity?		

9

Process Management in Practice I: Putting It All Together

The fundamentals of process management described in the preceding chapters can be implemented in seven simple steps to manage and improve any process, regardless of complexity. Managing a process is accomplished by defining work activities, making measurements at appropriate steps in the process, and taking corrective action on process deviations as they occur. The basic steps of process management are:

1. Determine process ownership.
2. Delineate process boundaries and interfaces.
3. Define the set of work activities in a clearly understandable way.
4. Determine customer requirements.
5. Establish control points.
6. Measure and assess the process.
7. Obtain feedback and perform corrective action.

The application of these steps can be illustrated by showing how they can be applied to a business situation. The following scenario frequently occurs in many firms having service-type or administrative operations and illustrates a relatively simple process functioning in a reactive mode.

Mr. Smith is a new manager of the XYZ department. Twice a year, in January and June, his department must formulate and send out 25 copies of an important company document, the Strategic Plan. The Strategic Plan contains certain material requiring a response back to Smith within two weeks after issuance. Since this document and the needed responses have high management visibility, any problems in this area can affect both the image of his department and Smith as a manager.

In early June, copies of the plan, received from duplicating services, were packaged by members of his department and sent through internal mail to each person on the distribution list kept by the department secretary. No problems with the mailing were reported to Smith. In mid-June, Smith began to realize that the requested responses were far short of what he expected—only three replied. He was embarrassed to report to his management that he could not provide a complete response to the document by the original input date. He immediately started making phone calls to the 22 people who did not respond and was surprised with the responses:

- 3 did not receive the document
- 2 just received the document at their new locations
- 7 complained of missing pages
- 4 reported pages out of sequence
- 5 reported that they could not respond in only 1 week
- 1 left the organization

The following day, Smith received a phone message to be in his manager's office the next morning to present his analysis of the problem and his action plans for preventing a reoccurrence.

Let us see how the seven basic steps can be applied to this situation.

Step 1: Establish Ownership

For any process, establishing and recognizing ownership is a necessary first step. In this case, Smith is actually the "owner" of the set of work activities within his department. Clearly, Mr. Smith is the manager of the department that originated and sent the document and is perceived by his manager as accountable for the problem of late and missing responses.

It is not clear, on the other hand, that he "owns" the total operation involving document distribution. In fact, as a first-line manager, he does not have control over duplicating services and the mail room—both of which are part of his process. However, Smith is responsible for a substantial part of the operation (origination and distribution) and is perceived both by his management and customers as the "owner." Smith has de facto responsibility for the set of activities extending from document origination through duplicating services, his own department, and the mail room. He is perceived as being accountable for the output quality of duplicating services as well as distribution even though these departments do not report to him—a common perception in processes crossing lines created by organizational boundaries.

Step 2: Establish Boundaries and Interfaces

Smith now needs to determine the process space within which he wants to examine this operation. Suppose he decides that he wants to look at the process from the point at which the original document is received at duplicating services to the point at which a copy is sent to the customer. This will set the boundary conditions. Figure 9-1 shows how boundaries may be denoted on a flowchart.

Establishing boundaries delineates the process space to be examined and facilitates a focus on a specific set of activities. By setting boundaries, we are also forced to think of the primary inputs and outputs into and out of the boundaries, who provides the inputs, and who receives the outputs. Boundary setting may also narrow down an ownership question in cases where it is an issue.

During this step, Smith should also note the key interfaces. Here we see that interfaces exist at the point of shipping the copies from

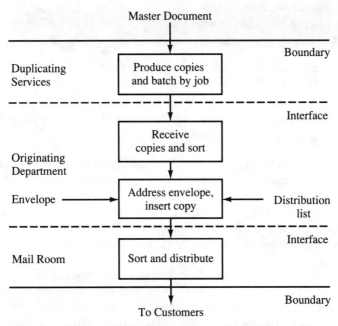

Figure 9-1.   Document distribution process: macro flowchart.

duplicating services to Smith's department and at the point of taking the packaged documents to the mail room. From the telephone responses Smith received, it is apparent that there are potential interface problems underlying this subprocess. As noted before, in managing a process, it has been found that typical problems affecting an operation lie at the interfaces. Denoting them assists in focusing on work flow requirements between the supplier and the recipient of the work product.

Step 3: Define the Process

Having bounded his process, Smith can proceed to define it. Typical of many administrative processes, Smith's is undefined. An accurate, understandable description of the process is fundamental in process management. Flow charting is the technique most often used. It provides a symbolic description of work activity and is relatively simple to use.

Typically, individuals will define a process from a personal perspective. More accurate and meaningful process descriptions are generally the result of a department team effort or the result of personal interviews with people performing activities (as noted in Chapter 7). Process definition is probably the most difficult step to complete because most people are unaccustomed to thinking of work-flow in terms of a relational set of activities.

Once having bounded the process space that is to be defined, Smith has a choice of:

- Getting very detailed and drawing a process flow of all tasks involved within the defined space, or

- Examining the process in terms of activities, then later detailing the flow at a task level when specific sets of activities seem to offer opportunity for improvement.

He chooses the latter course of action and lays out his process in terms of a macro or high level flow first. Where no process documentation exists, it is usually beneficial to first develop a macro level flowchart. This can be done relatively quickly and enables one to see the total process. From this one can develop an activity level flowchart. Frequently, process improvements become obvious as the activity flowchart is developed.

Smith starts the process flow diagram by drawing the input and output boundaries. Working from the input side, he notes the primary input that triggers the process. In this case, it is the original document that is brought to the reproduction services department for copying.

An instruction sheet is filled out and the original copied per the information on the instruction sheet. The copies are bundled and sent to the requestor. He then goes through the steps followed in his department: copies are inserted into envelopes, the envelopes are addressed, bundled together, and taken to the mail room. The mail room then distributes the envelopes to the departments where the individuals on the distribution list are located.

Figure 9-1 is a macro representation of the way Smith visualizes his process, showing the original document going to duplicating services, copies returned to the department for merging into envelopes, and envelopes subsequently taken to the mail room. Smith then can define his process at a more detailed level. He can do this readily because the macro diagram serves as a framework for it. Having defined his process, Smith is now ready to examine its control features. But before this is done, it is appropriate to determine customer requirements because this may affect subsequent process steps.

Step 4: Establish Customer Requirements

Smith has primary, or external, customers as well as internal customers. His primary customers are the 25 people on his distribution list. Typical of many administrative processes, the producer of the output has little or no idea of the requirements of the output as viewed by the customer. Smith resorts to a sample phone survey and finds, to his surprise, that customers not only require a perfect copy of the Strategic Plan but tell him they need two weeks to evaluate and prepare a reply to it. Smith now realizes that he has an ineffective process—the plan is distributed in an untimely fashion and is deficient to begin with.

He contacts the two internal customers: duplicating services and the mail room. The manager of duplicating services informed Smith that the original copy of the Plan had out-of-sequence pages to begin with and assumed, because it was a master copy, that it was in the correct sequence. In addition, the manager reported that 25 copies of a large document such as the Plan required a two-day lead time. Smith's secretary requested same-day service, not understanding the lead-time requirements. As a result, two and a half days elapsed in the copy center.

The supervisor of the mail room informed Smith that, for mail to be distributed the next day, envelopes must be brought directly to the mail room by noon. Envelopes that are picked up in his building require an additional day. Further, some of the people on his distribution list are situated at outlying locations where as much as four days are needed to receive mail. Smith begins to see that as much as a week elapsed before the Plan was received by his customers. At this point, Smith is not sure about all of the fixes needed but knows that something has to be done right away. The logical thing to do is to see where he can introduce some in-line checks and measures in his own area without getting into a hassle with his counterparts in duplicating services and the mail room.

Step 5: Establish Control Points

In Smith's process, no control points exist. There are no checks on the quality of the copies provided by reproduction services nor is there any verification of the accuracy of the distribution list. Smith's process is essentially lacking in any activities that could provide control.

As noted in Chapter 5, control points are steps in the process where a check, review, verification, or inspection is done. At this step, we have the opportunity to correct a defective input or return it to the person supplying it for correction. An example of a control point is a cashier reviewing and verifying an expense account for both correctness of the expenses and required account information. At this step, the cashier can return a deficient submission or perhaps correct the deficiency with the submitter of the expense account at the time of review.

As Smith looks at his flowchart and reviews the customer feedback he has received, he notes that seven complained of missing pages and four reported pages in the document that were out of sequence. It becomes apparent to him that he has to do something to assure that copies of the Plan received from duplicating services are correct and free of defects.

The first decision he makes is to have an inspection of the copies done in his department by his own people—a not uncommon action. With inspection implemented, he knows he will be able to eliminate, or at least reduce, the defects from this part of the process. However, he is expending resources to fix his supplier's quality problem and reducing the productivity of his process by introducing more labor for a given output.

On the other hand, if he were to employ the customer-producer-supplier model (Chapter 3), he would meet with his supplier and state his requirements for defect-free copies to be shipped to his department. Duplicating services, in turn, would, as a supplier, be obligated to review its own operation and make the appropriate changes to ensure that Smith's requirements are met. We note that, for the copying activity, Smith is both a customer and a supplier; he is a customer for 25 copies of the document as well as a supplier of the original document to be copied. Therefore, Smith must ensure that the original supplied to duplicating services is defect-free. He does this by making sure the department secretary checks the quality of the original prior to taking it to duplicating services.

Continuing to review the feedback, Smith sees that three people didn't receive the document at all, two just received it at their new locations, and five said they could not respond in only one week. He begins to realize that he also has problems with the proper addressing and timely mailing of the documents. He sees the opportunity of making additional process changes within his department to address these problems: having the distribution list checked for the most current addresses prior to addressing envelopes and having the list updated as part of the document distribution process. He also takes steps to have the date and time of mailing recorded so he knows when the documents were taken to the mail room.

Three control points have now been inroduced: checking the quality of the original, inspecting the copies received from duplicating services, and inspecting the distribution list. By introducing these points (see Figure 9-2) he has developed a means for controlling his process and is ready to examine the sixth aspect of process management, metrics and evaluation.

Step 6: Measure and Assess the Process

Proper measurements are the heart of control. Some practitioners find this the most difficult step in process management because the problem becomes one of what to measure. We often tend to think in terms of measuring defects or redoing a task within our operation. But if we look at each step, we can see other measures that govern the effectiveness of our process. Such things as response time of a request being processed through a department are measures that can be as important as a defect measure in assessing its effectiveness.

Let us suppose, for example, that our department is responsible for producing and distributing a document. The goal we have set is error-free output. To achieve this goal, we may spend so much time and resources in inspecting, correcting, and reinspecting documents that the timeliness of distributing them is lost. Our process, though it may be capable of achieving an error-free goal, is only partially effective, because the copy did not reach the customer in a timely fashion. In addition, we have lost efficiency in that we have applied additional resources to do inspection without addressing the root causes that are affecting the process. Without measurements, we are hard put to provide a factual basis for addressing these issues.

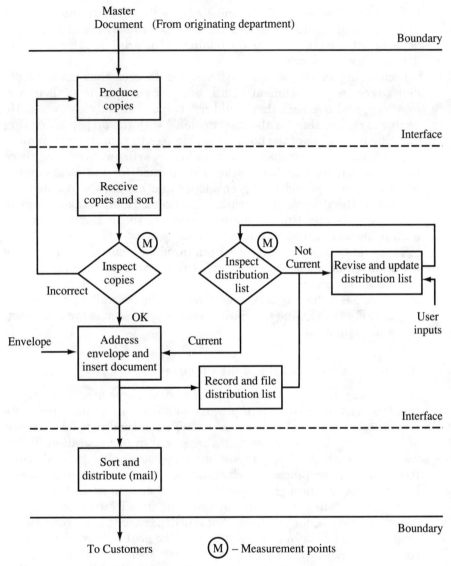

Figure 9-2.   Document distribution: Establish control and measurement points.

Let us examine how Smith can implement measurements. In Figure 9-2, we see where the copies received from duplicating services are inspected—a control point. This stage of the process provides an opportunity for measuring the quality of duplication. It is advantageous to record errors by defect categories, such as missing pages, out-

Data sheet				
Month	June	January	June	January
Period	1	2	3	4
Error Type				
Distribution	11	9	4	3
Reproduction	11	—	5	4
Typographical	—	8	3	2
Total	22	17	12	9

Check sheet				
Month	June	January	June	January
Period	1	2	3	4
Error Type				
Distribution	JHT JHT I	JHT IIII	IIII	III
Reproduction	JHT JHT I	—	JHT	IIII
Typographical	—	JHT III	III	II

Figure 9-3.   Data and check sheets for Mr. Smith's process

of-sequence pages, and so on, to facilitate root cause analysis and corrective action.

Data recording can be done using data sheets or check sheets in either hard copy or electronic form. Check sheets are known as one of the seven tools of quality.*

An illustration of sheets that Smith can use is shown in Figure 9-3. These sheets are simply tables arranged to allow recording of various attribute data taken on quality characteristics. They are also a means for classifying and organizing data to be taken for subsequent statistical analysis.

Examining Smith's findings for period 1, we see that, of the 22 reported problems, 11, or 50 percent, were related to copy quality—a likely area for improvement. The remaining could be categorized as distribution errors caused by either improper addressing or mailing. Using the distribution list control point, data regarding the frequency and type of changes to this list can be taken and monitored to determine the stability of this part of the process. A third measure that is appropriate for this process is typographical errors. This is shown on the data and check sheets of Figure 9-3.

---

*See, for example, the article in *Quality Progress*, October, 1990, "The Tools of Quality, Part V: Check Sheets" for a detailed description of check sheets. Check sheets, in contrast to data sheets, allow one to interpret results from the form in terms of a bar graph or histogram.

All three measures—distribution, reproduction, and typographical—reflect customer requirements. Smith is now ready to plot and track the performance of his department's output and take the seventh and final step of process management: feedback and corrective action.

Step 7: Perform Feedback and Corrective Action

Providing feedback to the participants in the process and taking corrective action on untoward process deviations is of fundamental importance in both stabilizing and improving the process. By doing two things—charting the measurements and setting targets—we can compare actual performance of the process against a standard, or norm. When we see actuals exceeding our targets, we need to take action to correct this condition in order to manage the process. In some cases, corrective action is simple and straightforward; in others, it is more complicated and drawn-out. In any case, without management attention to process performance and without corrective action, processes will decline in effectiveness, efficiency, and quality.

Smith has decided to use the trend and bar charts of the type shown in Figure 9-4 as a graphical means to display process performance. Note that he has decided to extend his measurements to include typing quality as a way to track internal quality performance. In the case of trend charts, he may consider establishing sliding or ratcheting targets (as described in Chapter 5) to reflect a continuous drive for process improvement. This is also illustrated in the trend chart of Figure 9-4 and shows the results of Smith's improvement actions in subsequent time periods.

Let us now review where Smith stands in light of the information he received and the actions to be taken to prevent a recurrence of the problem:

- The department secretary, as a standard procedure, is instructed to check each finished original for typographical errors and record errors prior to carrying the original to duplicating services. Significant time is provided to ensure that its lead time can be met.

- If inspecting the next set of documents again shows missing, illegible, and out-of-sequence pages, he will identify the defects to duplicating services and either return the documents found defective or reject the entire job. If enough time is provided for rejecting the job and having duplicating services redo it, he will find this to be a more effective way of getting results compared with reworking the documents in his own department.

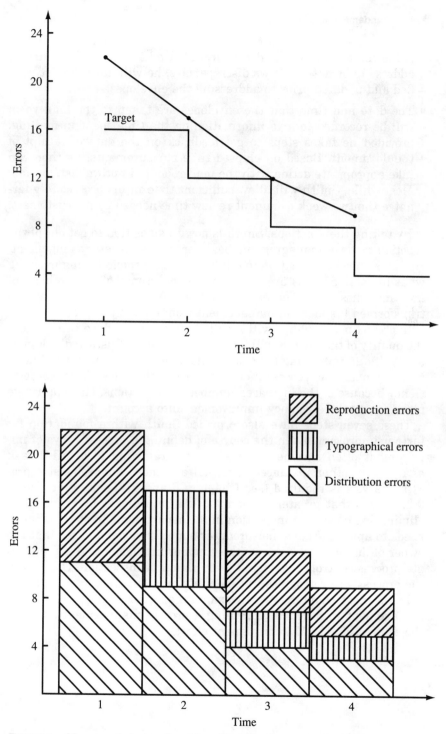

Figure 9-4. Measurements: trend and bar charts for document distribution.

- If the comparison of the distribution list addresses against current address information shows discrepancies, he will have the list verified and updated prior to addressing the envelopes.

- The date and time that the envelopes are taken to the mail room will be recorded to give information on the time spent in mailing, provided he takes steps to get feedback on the date of receipt of the document. Based on elapsed time measurements, he then can take appropriate action with the mail room to improve distribution. Meanwhile, Smith will allow sufficient time in his process to guarantee the two-week document review time needed by his customers.

By taking these actions, Smith is now in a position to get out of the reaction mode of managing and begin proactive process management. He is also now in a position to tell his manager the following morning what his action plans are to prevent a recurrence of the problem of late and missing responses. By taking these steps, Smith has also transformed his mode of management from reaction to prevention.

Improved performance will result. Defects will reduce in time and the quality of the process will improve. The trends illustrated in Figure 9-4 typify an improving operation. The actions that Smith has taken, however, make his process more effective at the price of reduced efficiency because of the increased number of inspections. His long-range plans must take efficiency improvement into account.

These seven steps have also provided Smith with a foundation for further improvement. In the course of defining a process, ways of improving it often become obvious without resorting to detailed task analysis. In other instances, a formal and more comprehensive process analysis is required (see Chapters 6 and 7). Looking at Figure 9-2, we see that greater efficiency could be achieved by reducing or eliminating document inspection in the department. To do this, Smith needs to apply the customer-producer-supplier model and see that the owner of duplicating services (the supplier) implements the appropriate process controls to warrant elimination of inspection. Otherwise, the process will retrogress in time to its previous state.

In the next chapter, we will examine another example of process management in practice: launching a process management study in the context of a TQM improvement effort.

10

Process Management in Practice II: Implementing Process Management for TQM

In the previous chapter, we examined the application of process management to a process operating in a reactive mode. Inputs to a planning process were both late and missing because the outputs of the document distribution subprocess were deficient in quality and timing. Process management was applied reactively in this case to a subprocess that was essentially broken or malfunctioning. In this chapter, we examine process management in terms of applying it proactively to support a TQM improvement effort.

TQM and Its Features

The background of total quality management (TQM) was described in Chapter 1. Interest in quality improvement in general, and TQM in particular, has exploded in recent years. Improvement as a formalized, directed activity has become widely publicized in the last decade. Many organizations have embarked on quality improvement with varying degrees of success. Numerous approaches have been used, depending on the school of quality philosophy influencing management, experiences of other organizations, the influence of consultants, and the influence on management of touted panaceas such as SPC and quality circles.

The lack of a systematic improvement philosophy or vision is prevalent in many U.S. companies. Masaki Imai, in his book *Kaizen*, describes a fundamental difference in the approach that Japan and the U.S. take in managing operations. In the U.S., improvement is neither a part of the day-to-day operations of a firm nor its tactical strategy whereas, in Japan, it is ingrained. A western manager perceives his or her job to be mainly one of maintenance or tactical activity, whereas the Japanese manager views improvement as a basic managerial responsibility.

There is no single, unique approach to improvement that works to the exclusion of others. It has been found, however, that successful and sustained improvement requires, first, ongoing commitment and involvement by management. Second, it requires a quality philosophy and a strategy consistent with organization culture and values. Third, it requires continuous motivation and participation by employees and, fourth, permanent improvement requires that a system be put in place to sustain it on an ongoing basis. TQM contains these features.

TQM, as we know it today, does not provide a methodology or prescription for improvement, but understanding its features can lead one to deduce certain requirements for successful implementation. Its main features are:

- Management leadership, commitment, and participation
- Participation of all functions of the organization
- Employee commitment and involvement
- Customer orientation
- A system for continuous improvement
- A means for assessing progress
- Training
- Recognition
- Communication
- Strategy and deployment

Conditions of excellence

Some of these features have been embodied by firms in what are called conditions of excellence. In the early 1980s, Westinghouse developed 10 conditions of excellence for total quality; in 1985, they were expanded to 12. These 12 conditions are divided into four major areas:

1. Customer orientation
2. Human resource excellence
   a. Participation—employee involvement
   b. Development
   c. Motivation

3. Product/process leadership
   a. Product/service excellence
   b. Processes/procedures
   c. Information systems
   d. Suppliers
4. Management leadership
   a. Total quality culture
   b. Annual improvement plan
   c. Communications
   d. Accountability structures

With the exceptions of recognition, strategy, and deployment, virtually all of the main features of TQM are embedded in these conditions. At Westinghouse, total quality is measured by three criteria: value/price ratio as a measure of customer satisfaction, value/cost ratio as a measure of financial success, and error-free performance as measured by the cost of nonconformance—a cost of quality measure.

Based on IBM's work in the 1980s, R. F. Boedecker[1,2] described 11 conditions of excellence for total quality improvement:

1. Management action, commitment, and involvement

2. Policies, strategies, and plans for improvement

3. An organization for implementing improvement

4. Education

5. Measurements and committed improvements

6. Regular reviews of plans and progress

7. Rewards and recognition

8. Participative management methods

9. Communication

10. Supplier quality improvement

11. Process management of the business processes

As can be seen, many of the basic features of TQM are also embodied in these conditions. One of the conditions, supplier quality improvement, is specific to product organizations where supplier quality is integral to the quality of the final product.

Management's role

Because TQM represents a change in the manner in which organizations conduct business, managers are the primary agents in promulgating change. The command structure of many organizations suggest that upper management must provide leadership if change is to occur. This does not mean that the CEO or president of a firm must be the only initiator of change. Often, in decentralized organizations, TQM is initiated at management levels below the CEO and, after a period of time during which awareness and realization occur, top management may assume leadership. IBM, Ford, General Motors, and AT&T are examples of this. In higher education, both types of change initiation are occurring. At universities such as Rhode Island and Samford and colleges such as Marist, presidents initiate and promote TQM. At other institutions like the University of Oregon and the University of Pennsylvania, TQM has been initiated by a vice president. Less frequently do we find faculty initiating it—Fordham University being a case in point.

As with any new philosophy or way of thinking, TQM is subject to various degrees of acceptance, ranging from enthusiasm to complete rejection. Resistance to change and cultural conflicts within the organization are factors that must be understood. For this reason, leadership, demonstrated commitment, and involvement by management are necessary to facilitate and maintain the change process. Too often, inconsistent management behavior is construed by employees as a lack of conviction.* The old adage, "actions speak louder than words" and the newer phrase "walk the talk" is apropos in TQM. Successful TQM efforts have consistently shown that leadership, support, and involvement by all levels of management exist. On the other hand, it can be shown that failed or flagging TQM activities are largely attributable to lack of interest or involvement by management, particularly middle managers.

Leadership, support, and involvement comprise such activities as reviewing the status of individual and team efforts; providing encouragement, praise, and advice where appropriate; participating in or chairing a quality council or steering committee; establishing strategic direction; providing recognition and rewards; setting an example; and communicating both verbally and in writing.

Organization-wide involvement

By definition, an organization successfully engaged in total quality management can also demonstrate that all functions of the organization are participating. This is a manifestation of the systems approach

---

*One of Deming's 14 points, "Constancy of purpose," implies establishing direction and maintaining it. When employees react to a new program or direction by management by saying "It's another fad that will go away," they are responding to an inconsistency that they perceive in management.

to quality management and means the total involvement of an organization that produces a product or service. If, for example, one looks at a product firm, coordinated quality improvement efforts in manufacturing, product development, procurement, field service, finance, and marketing should be readily apparent if it is engaged in TQM. In addition, one should be able to see supporting functions such as information systems, human resources, facilities, and so on involved in similar improvement efforts.

Employee involvement

It should be self-evident that TQM requires both commitment and participation of all members of an organization. In firms, employees are the crucial factor in quality improvement. Management must ensure that employee buy-in to TQM occurs because improvement is developed and implemented by the employees. Because of the multifaceted nature of work activity, much of the improvement effort is conducted by teams. Quality circles, empowered teams, and quality improvement teams (QIT), are examples of this. The team approach implies recognition of the fact that the whole is greater than the sum of its parts.

There are three types of work teams that have evolved in the last 30 years:

- Problem-solving teams
- Task-oriented or special-purpose teams
- Empowered or self-managing teams

Problem-solving teams are based on the Japanese quality circle group approach. The concept was brought into the United States in the late 1970s primarily through the work of Wayne Rieker and grew rapidly during the 1980s as an initial effort to achieve quality improvement. Problem-solving teams usually consist of groups of four to eight employees from the same department who participate in solving and improving work-related problems usually during working hours. Many of these problems relate to the quality and productivity of work. Although the results of these teams have been significant in a number of instances, many achieved only minor and easily accomplished improvements. Few have managed to survive for any length of time. Lack of power to implement changes and lack of sustained support and interest by management are two of the main causes affecting the survival of these teams. Quality circles have been termed a management fad.[3]

Task-oriented employee teams, sometimes known as task forces or "tiger" teams, have existed for some time. IBM, for example, has used this concept since the 1950s on a selected basis. Employees are chosen by management according to skill and knowledge and assigned to a task team to solve a specific problem. They may be empowered to implement their solutions or improvement recommendations. Task teams have a defined purpose or objective and exist for the time it takes to accomplish the objective. Often, team members are assigned full time to work on the team, and the team is given supporting resources needed to accomplish its objective. Case 2 described in Chapter 11 is an example of a task team assembled for purposes of improving a labor-claiming process.

Empowerment is a form of the job enrichment concept popularized in the 1960s and 1970s in which jobs are designed from a vertical perspective, providing a worker with more decision-making responsibility. Together with job enlargement, which is task broadening, these concepts provided management with a means to improving productivity through job design. Empowered, or self-directed, teams represent current management thinking in regard to the employee role in the workplace.

In manufacturing, this approach involves work teams of five to ten people to produce an entire product whether it is an end item or a subunit. The team is fully responsible for setting work and off-time schedules, ordering materials, and solving line problems—responsibilities previously reserved for supervisory management. Members are cross-trained and are required to be familiar with all job tasks. They are selected for their ability to work as team members. Volvo, the Swedish auto manufacturer, is one of the pioneers of this concept. Significant improvements in productivity—30 percent and greater—have been reported.[4] Examples are given in Chapter 15.

Customer focus

Another feature of TQM is a customer orientation or focus. For products and services, the revenue-paying customer is the reason for the firm's existence. Without a customer purchasing a product or service, the organization cannot exist. TQM not only reminds us of this fact but also implies that quality management must continually take into account the determination and satisfaction of customer needs.* For some

---

*A TQM practitioner should also be knowledgeable of the Kano model of customer satisfaction. In this model, the aim of a producer of a product or service should be to provide "exciting quality," that is, a quality level that delights the customer and transcends merely meeting customer requirements. In other words, the objective of producers should be to exceed customer expectations. Producers who have done this have obtained a competitive advantage and have increased market share. Certain Japanese consumer products such as luxury autos exemplify a trend to exceed expectations.

organizations, such as IBM and Federal Express, customer focus has become the primary element driving their TQM efforts. IBM, for example, has renamed their program "Market Driven Quality."

TQM also involves adopting an internal customer focus. As a result, business operations should be examined and improved by taking into account internal customer needs. It is implied that the needs of internal customers are ultimately a reflection of requirements of the external customer. The customer-producer-supplier relationship model described in Chapter 3 is useful for understanding customer requirements in terms of the producer's and supplier's capability. Also useful are deployment matrices and other means of documenting and translating requirements described in the same chapter.

Continuous improvement

A system or mechanism for continuous improvement is another essential feature of TQM. Once an improvement is achieved, a mechanism must be in place to enable improvement to continue. Quality improvement is viewed by practitioners as a journey—a never-ending process. There are various methods for achieving this. The concept of annual target setting first proposed by Juran is one such method. Here, a target quality level is set for a product, a service, or an operation at the beginning of the year. Improvements are made and measurements performed and compared with target. By the end of the year, the improvements made are intended to achieve target. If it is achieved, a new target is set for the following year and the improvement process begins over again.

In this manner, the Japanese have been able to achieve unprecedented levels of quality in their consumer products and have set quality standards for the rest of the world. They have embodied this philosophy in the term *kaizen*. In contrast, the philosophy of continuous improvement is conspicuously absent in much of U.S. industry, a primary reason being that managers tend to be measured on short-term parameters such as cost, revenue, and profits rather than quality improvement, which is a longer-term effort.

Other continuous improvement methods involve the notion of the "ratcheting" target and the zero-defect goal approach. Similar to the yearly target method, ratcheting involves setting a quality target,* improving the product to achieve the target, and then resetting the target

---

*Target setting can be both statistically derived based on process or design capability and set by management judgment in terms of the potential that can be achieved. It has sometimes been set by management decree in a somewhat arbitrary fashion.

to a higher quality level. The new level is management-set and is aggressive but usually not unreasonable or irrational. There is no fixed interval for resetting targets. It is dependent on the natural forces occurring within the improvement process and the nature of the product or service. IBM has employed this method in mainframe computer and component production; its origins were described in Chapter 5.

The zero-defect method involves setting an objective of zero defects initially. Interim targets are set to provide realistic and reasonable levels that can be achieved. Improvement programs are put in place to achieve these levels. This approach has prompted heated debate as to its practicality as a meaningful goal or target. A modification of this method is the Six Sigma Program of Motorola.[5] Six Sigma was derived in part from the properties of the normal distribution and essentially means a target defect level of 3.4 parts per million or a quality level of 99.999997%. The far-reaching aspect of this target is intended to promote continuous improvement to achieve this level. In this context, Six Sigma can be viewed as a goal.

Assessing progress

A corollary to having a system for continuous improvement is providing a means for measuring progress. Having goals or targets in place is a necessary first step; measuring improvement is a necessary second step. Chapter 5 discusses various types of measures that can be employed in assessing improvement. Measures are both external and internal.

The primary external measures are customer satisfaction, cycle or response time, and defects or errors. Because customer focus is a key element in TQM, satisfaction measures become important in gaging product and service quality improvement. The widely publicized J. D. Powers survey of consumer satisfaction with cars is an example of this. Internal measures are primarily process measures such as activity timing, errors, and the number of recycles.

In addition to having measures, it is important to have a system of management reviews for assessing progress. Management reviews act as a forcing function for accelerating activity, getting problems resolved, making measurements, and evaluating status. Without a consistent, well-formulated and regularly scheduled review process, TQM will, in many organizations, receive a low priority in the daily activity schedule and, consequently, will either be forgotten or postponed. Reviews also show management's involvement and participation in TQM as well as its interest.

Training

Training of managers and workers in both job-specific and quality-related matters is an integral part of TQM. Job-related training has been found to be an excellent investment for obtaining a productive work force and quality products and services. Quality-related training for both management and employees has been found to be essential in TQM and ranges from courses in participative management and team building to quality tools and problem-solving methods. Training is a key element of a successful TQM implementation strategy.

TQM training is generally laid out in the following sequence.

Topic	Duration (Days)
Quality awareness	$\frac{1}{2}$–1
TQM-specific training	2–5
Tools training	1–5
Team behavior	1–3

Quality awareness training comprises an introduction to TQM, rationale, definitions, organization strategy and direction, and management expectations of the employee. TQM-specific training involves training in the methodology to be used, such as project or process methodology or a multiple-step prescriptive approach (such as the Crosby 14-point program). Next is tools training, which involves techniques such as problem-solving models; cause-effect diagrams; brain-storming; Pareto diagrams; various statistical methods, ranging from elementary graphical tools to SPC; Taguchi methods; and quality function deployment (QFD), which was mentioned in Chapter 3. Tools training may be divided into specific topical areas or grouped into elementary, intermediate, and advanced courses. Finally, team behavior training may either be reserved for team leaders and facilitatiors or given to all participants. Topics include practice exercises in team building, understanding group behavior, and communications.

Recognition

Recognition has been found to be an important motivator for sustaining continuous improvement. There are no standards or prescriptions for management in recognizing individuals or teams involved in TQM and accomplishing significant results. Recognition can be tangible as well as intangible and takes the form of both peer and management cognizance of contribution. It can be either monetary or symbolic in nature. Awards that organizations use (both singly and in combination) are cash, a check to cover a dinner for two, U.S. savings bonds, embossed pens, engraved plaques, framed certificates of accomplishment, and symbolic mementos such as pewter elks (as in the case of Hartford Insurance) as well as luncheons and dinners.

The appropriateness of a specific type of recognition varies from one organization to another and even within an organization. Culture and values, wage and salary levels, employee needs, individual and team perceptions of reward, and equity of awards are some of the factors that should be considered in developing a recognition program. Recognition should be timely and given as close as possible to the time at which the improvement effort was completed. Recognizing accomplishment long after a team has finished its work is ineffective and can be construed as lack of responsiveness and interest by management.

Communication

For any concept that represents a change in organizational behavior, communication is vital. It is no different with TQM. It is important that all key aspects of TQM (management and employee commitment, customer focus, improvement, recognition, and strategy or direction) as well as ongoing status and achievement be communicated to the participants of the organization, both management and employees. Communication is accomplished by newsletters, special memoranda, group meetings, video conferencing, videotapes, auditorium meetings, and simple printed messages.

Since TQM represents a change in the way an organization conducts business, it is essential that the rationale for the change and the strategy for bringing it about be communicated to everyone so that acceptance or "buy-in" can occur. It is also necessary so that everyone knows the direction that has been established for the organization and the reasons for it.

Strategy and deployment

A strategy is essentially a game plan for doing things and, therefore, a means for transforming TQM policy, philosophy, and methods into action within the organization. This is known as deployment. Deployment is inherent in a well-executed TQM strategy. It is also implicit in the Baldrige Award criteria for examining a quality system. In some organizations, particularly industrial, strategy is established by upper management. In others, it is a result of the consensus of participants after a sufficient period of discussion and debate.

A typical TQM strategy for an organization may contain the following elements:

- A preface including a definition of TQM and the reason for introducing it to the organization.

- The role of TQM in relation to the basic mission and purpose of the organization.

- A description of the components of TQM as seen by the organization.

- A set of goals.

- A description of the method of implementation of the TQM components within the organization. (For example, if training is a TQM component, a general description of the type of management and employee training to be undertaken would be provided in this element.)

- A description of how the strategy is to be deployed within the organization.

- An overall time horizon for the strategy.

Improvement Through Process Management

Quality improvement in the past has been approached in several ways. One popular approach involves the selection of improvement opportunities or projects within an operation—often by a group or team. Selection is often based on operational problems perceived by the group or by management. Following this, problem analysis, root cause determination, and corrective action are undertaken. Results are then measured. This approach is generally problem-specific and has been termed "firefighting." Rarely is the broader issue of suboptimization or the concept of total quality addressed by a problem-specific methodology.

Other approaches that have been popularized involve adopting a multiple-point program consisting of dictums for management or prescriptive steps. These were described in Chapter 1.

The most important features of TQM are its customer orientation, its emphasis on continuous improvement, and its organization-wide aspects. Process management provides a means for addressing these features and serves as a unifying methodology for TQM.[6] If we examine the output of an organization such as a for-profit enterprise, we see that this output is the result of a set of interrelated business processes. Process management is a methodology for addressing the improvement of these processes. Embedded in this methodology are the TQM features of customer orientation, assessment, and continuous improvement, and an emphasis on prevention. The latter is sometimes referred to as "quality at the source."

Adopting a process approach in a TQM framework involves the following steps:

1. Select the processes.
2. Develop an improvement plan.
3. Form the process teams.
4. Perform a process management analysis.
5. Define the improvements—both short- and long-term.
6. Obtain owner approval to implement improvements.
7. Assess effectiveness of the improvements.

1. Process selection

Selecting and choosing business processes for improvement can be done on the basis of one or more of the following criteria:

- Management judgment and preference
- Customer complaints
- Employee complaints
- Perceived importance and criticality
- Effect on customer satisfaction
- Random selection
- Process size and cost

The first five are essentially subjective, judgmental, or reactive in nature and need no elaboration. On the other hand, a process can be selected based on its size and operational cost. Implied in *size and cost* is improvement opportunity—the larger the process, the greater the chance that redundant and other non-value add activities are present, which affords opportunity for their elimination. A method known as system disaggregation provides a way of enumerating and positioning the processes contained within an organization and serves as a basis for selection based on the other criteria.

System disaggregation. With any business entity, it is useful at the beginning to develop a picture or model of its productive system to see how various processes compose the total organization. A useful means for disaggregating a productive system into its key processes is the value chain model.[7] The value chain, first proposed by McKinsey and Company, is based on the fact that a firm consists of a "collection of activities" or business processes that exist to develop, manufacture, sell, and distribute a product or service. An organization's business processes can be mapped by using a value chain.

This model is based on the premise that a firm's "activities"[*] can be classified into ones that create a product or service of economic value to customers (called primary activities) and activities used to support them (support activities). For a product type of firm, there are six primary activities: product development, inbound logistics, operations, outbound logistics, marketing and sales, and service (see Figure 10-1).

Product development comprises the design, prototyping, design validation, and release to manufacturing of the product. A macro level flow of this process can be found in Chapter 13, Figure 13-1. Inbound logistics comprise warehousing and other materials management activities. Operations involves the actual producing of the product or service. Outbound logistics comprises finished goods, warehousing, order processing, scheduling, and distribution. Marketing and sales as well as service are self-explanatory. These are also considered primary business processes because they provide value-add in terms of facilitating sale and transfer of a product to a customer as well as assistance after the sale.

Support or secondary activities in this model comprise such functions as finance, information systems, personnel, and quality control. These do not contribute directly to the value-add of the product or service but are required in the creation and sale of a product or are needed for control purposes.

Disaggregating a large organization into its primary and secondary components serves to provide a means for process identification and choice. An overall system perspective of the organization is obtained and its critical elements (process) identified. In its literal interpretation, TQM implies that *all* major processes of the productive system are identified and addressed. In practice, these processes are addressed in accordance with a strategy.

2. Improvement plan

Having selected the processes to be analyzed for improvement, developing a plan of attack is the next order of business. The larger

---

[*]Porter (reference 7) and McKinsey use the term "activity" to mean a major business process.

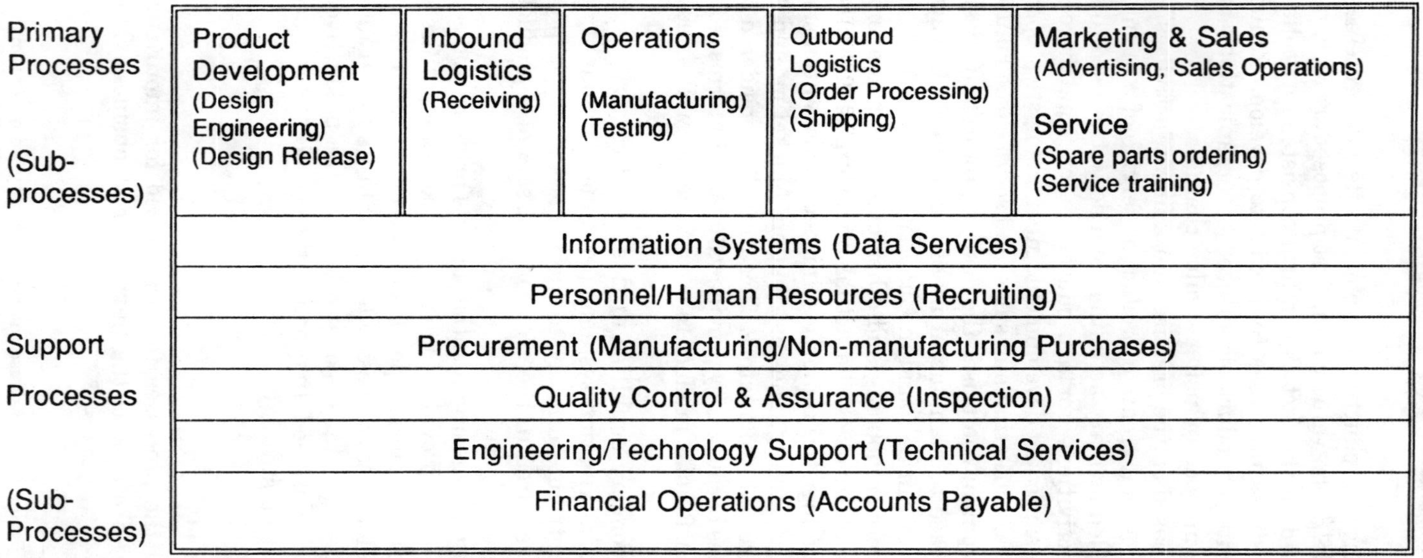

Primary Processes / (Sub-processes)

Product Development	Inbound Logistics	Operations	Outbound Logistics	Marketing & Sales
(Design Engineering) (Design Release)	(Receiving)	(Manufacturing) (Testing)	(Order Processing) (Shipping)	(Advertising, Sales Operations)
				Service (Spare parts ordering) (Service training)

Support Processes / (Sub-Processes)

- Information Systems (Data Services)
- Personnel/Human Resources (Recruiting)
- Procurement (Manufacturing/Non-manufacturing Purchases)
- Quality Control & Assurance (Inspection)
- Engineering/Technology Support (Technical Services)
- Financial Operations (Accounts Payable)

Figure 10-1. Value chain for a product firm. Sub-processes are indicated in parentheses.

the organization, the greater the need for a coherent, well-conceived plan.

The elements of a process improvement plan include the mission and objectives of the team, the type of training required, composition and staffing of the teams, support required (such as information systems), team meeting schedules, management review schedules, team progress documentation, and an overall process improvement schedule.

Where time and human resources become a constraint, it is appropriate for management to develop a priority or phase-in plan. In a three-phase plan, we first select the most important processes for phase 1 of the total improvement plan and develop a sequence of actions involving steps 3 through 7 for these processes. As the phase 1 processes are under way, we select the next priority set of processes for phase 2 and time the start of steps 3 through 7 for these processes so as not to impact the resources being applied to the phase 1 improvement activity.

Finally, once the phase 2 processes are under way, we select the least critical of all and launch steps 3 through 7 in a manner so as not to impact the work in phase 2. A phase-in plan such as this does extend the overall time span for improvement but minimizes the impact of resources on the organization.

3. Process teams

Because of the multidisciplined aspects of an operation, improvement is best accomplished by means of teams. Process teams can be either departmental or cross-functional, depending on the nature of the process. Team members should be selected not only for their knowledge and ability to contribute to specific areas of the process but for their ability to get along with other members. For certain processes, it has been found useful for the team to have members representing internal customers as well as suppliers to the process. In some cases, external customers are appropriate.

Each group should have a team leader either appointed by management or elected by the team members. The team leader should have knowledge of group dynamics, an ability to handle interpersonal relationships, and an ability to conduct effective meetings. In many cases, the department supervisor or group leader assumes the role of team leader and serves as a spokesperson for the group. In some cases, the process owner is the accepted team leader. In others, the owner is perceived to be an inhibitor and would not, therefore, function as an effective team leader.

Some teams may also require the services of a facilitator. Although it is not mandatory, facilitation appears desirable in some instances where the team leader does not exercise this function. The facilitator should also have some understanding of group dynamics, interpersonal relationships, and conducting meetings. The facilitator performs general coordination, sets agendas for the meetings, and, in general, supports the team leader. Some organizations have found it helpful to have a facilitator from outside the organization. An external facilitator can provide an unbiased, third-party view of the subject under discussion, which is often valuable in the development of the team. At the onset, all team members should have received process management training, including the team leader and facilitator.

Because the team activity represents a significant expenditure of human resources for the organization over a period of time and a change in the employees' work habits and priority, an explanation of the importance and the reasons why this activity is being conducted should be given by management. In some organizations, a kickoff meeting initiated by management (usually, the process owners) has been found useful in order to set a tone and describe the purpose, objectives, and expectations of the team. It also affords an opportunity for dialogue between the team members and management on any questions or concerns that members may have.

4. Process management analysis

Once launched, the process team can now embark on performing a process management analysis. The seven steps described in the previous chapter can be followed to accomplish the analysis, namely, ownership establishment, boundary setting, process definition, customer requirements definition, controls and measures, and assessment. During the analysis phase, it is important for the team leader and facilitator to ensure that a questioning attitude is developed as to why certain activities exist or are performed in the manner described. The answers often lead to the elimination or modification of these activities.

5. Process improvement

Both during and after the analysis, ideas for process improvement are generated. Improvements will generally take two forms: short-term or incremental and long-term or breakthrough improvements. Incremental improvements are the kind that provide some immediate relief or fix to a problem, require little or no capital investment, and can be accomplished relatively quickly. On the other hand, breakthrough improvements are innovative and long-term, create profound changes to the operation, and usually require capital investment. These are in the category of "re-engineered" processes.* Both types of improvement require the involvement and approval of the process owner. Breakthrough improvements will require sensitivity by management to the effects of reductions in staffing on retraining and reassignment of people. Examples of both incremental and breakthrough improvements are described in Chapters 11 through 13.

6. Owner approval and implementation

Upon completion of the improvement phase, the team should review proposed improvements and obtain approval of the process changes with the owner, after which the implementation phase can begin. The implementation phase is the longest of all. Its duration is generally in terms of weeks or months for incremental changes and months to years for breakthrough changes, depending on the complexity of the changes and resources available for implementation. The same team may be used for implementation or, depending on the skills required, another team may be constituted.

7. Assessment

The final step is process assessment. Once implementation is complete, it is natural to assess and measure the benefits of the improvements. Often, implementation and assessment provide a springboard for further improvement and the cycle repeats itself. In this way, the process approach provides a means for continuous improvement. Chapter 8 describes the various methods and criteria for process assessment and evaluation. In the following section, we will examine how the process approach to TQM has been applied in major organizations.

Examples of Process Management Implementation

Within the last five years, a number of major organizations have adopted and implemented process management in a manner similar to that described in this book. In this section, we shall describe versions that leading companies are using for managing business processes. The first is the five-dimensional model developed at the National Cash

---

*"Reengineering" is in current vogue in the United States. Much of this concept involves process simplification through the use of information technology and nontraditional or innovative ways of transacting business such as invoiceless payables.

Register Corporation. The second is the four-stage, process-management model developed by AT&T.*

NCR's five-dimension model

In the NCR view, "managing a process for continuous improvement requires that the process move through five 'dimensions' toward the goal of defect-free output".[8] These dimensions are:

- Ownership
- Definition
- Measurement
- Control
- Improvement

Each of these dimensions has been explained in earlier chapters in terms of the steps of process management. The NCR approach is based on an input-task–output-concept similar to the input-process–output-transformation model described in Chapter 2. It also includes the concept of the customer-supplier relationship described in Chapter 3. Improvement is implemented in the following manner.

First, a process is selected for improvement. A quality management team (called QMT) is then set up to identify and prioritize opportunities for improvement based on either strategic importance to the department, level of quality costs, ease of improvement, or the size of the problem. After selection, a quality improvement team (QIT) is organized and set up by the process owner to examine the process in terms of the five dimensions noted above.

The following example (excerpted from reference 8) of the NCR model is applied to mail services at corporate headquarters in Dayton, Ohio. This department is managed by Marj Lawson, a member of the quality management team.

OWNERSHIP

The Quality Management Team set out to manage the mail service process. The team began with the dimension of ownership. Marj assigned a manager as the owner of the process chosen to be improved. The owner

---

*It should be noted that these two models were developed independently by the two companies prior to the buyout of NCR by AT&T and are based on the IBM business process approach.

immediately started the culture change by sending the people in the process to process management classes and, subsequent to their training, asked their opinions on process improvements. Considering that many of these people had never participated in NCR education programs, this display of management involvement was widely accepted as a genuine concern for the employees and the work that was being done. This, in turn, sparked greater interest in the *employees* toward the work being done. A Quality Improvement Team (QIT) was organized by the owner to work specifically with this process through the dimensions of process management. The people were actually given time to stop and think about what they were doing and to think about improvements to their work activities. Now that the people were interested in improving the process, the definition phase could begin.

DEFINITION

Determining exactly what this work activity looked like and how all the pieces were or were not interrelated was a difficult task. It seemed as if everything was one big process. These processes were finally separated by looking at each class of incoming mail and each method of processing outgoing mail.

Determining who were the customers and suppliers was also difficult because it seemed as if everyone was a customer and, worse yet, these customers were also suppliers to that same process. The QIT finally determined that the customer was simply everyone who receives mail and the input is all types of mail. This allowed the team to focus on the purpose of the process: To deliver accurately all mail in a timely manner to the Dayton campus. Each type of mail had been handled separately, thus creating several very similar processes, such as overnight express courier "mail," first-class mail, consolidated "flats" from remote NCR locations, advertising pieces, etc. Once the process was viewed more simply as mail being delivered to NCR employees, it was much easier to understand what was in and what was not in the process. ...

Once the team began to evaluate the customer concerns being received, it became obvious that deliveries by overnight couriers were viewed as "mail" by the customer but it traveled by the "package process" (shipped services) to be delivered to the addressee. Employees with concerns described overnight shipments as urgent "mail" that was not delivered by 10:30 A.M. as promised by the carrier. The customer viewed the delivery as "mail" and the process owners viewed it as a shipment.

In addition, the requirement of the internal customer was determined (through surveys and interviews) to be that incoming overnight deliveries must be on the desk of the addressee by 10:30 A.M. the morning of delivery. In reality, however, deliveries were being made to Mailing Services prior to 10:30 A.M. and distribution throughout the campus put these deliveries on the addressees' desks in the afternoon. The requirements of the customers for overnight mail delivery were clearly not being met.

MEASUREMENT

As improvement and corrective ideas began to flow, it was obvious that the QIT had to know more about the operation before making major changes. Initial measures were put in place simply to find out the size of the work load and how it flowed. Examples of these measures are:

- Number of items received daily
- Number of items mailed daily
- Number of unprocessed items at end of day
- Number of complaints about service (accuracy and timeliness)

Other measures were put in place to try to determine exactly how well the current process was meeting customer requirements. Examples of these measures are:

- Length of time from intracampus sender to addressee
- Number of missorts
- Number of late overnight deliveries to addressee

These measures were collected initially and the work teams were given the opportunity to set their improvement objectives. The measures were made into charts and placed on the wall in the work area. These charts are now updated monthly. The process work teams collect and report on the measures themselves. Results, improvement ideas, and new goals are reviewed and rewarded at weekly staff meetings of department employees.

CONTROL

Each process must operate in a controlled fashion before it can be improved. A controlled process has consistent output and that output must meet the customer requirements. An improved process does the same but in a manner that is more efficient, contains less cost of quality, and is more effective. An improved process actually has a different definition than the original process. The control dimension is essential to process improvement because without it, changes and "improvements" could be implemented that would be directed at something other than the root causes of the problems.

This QIT used their measurements to help find the root causes and make permanent corrections to the process. The initial productivity measured gave Marj the lever she needed to present new requirements to one of her suppliers, namely, Federal Express. It was determined that 72 percent of the overnight courier envelopes were for ultimate delivery to one building in the campus—not the one to which current deliveries were made. All envelopes, tubes, and packages were delivered unsorted to one destination and then had to be resorted by building before delivery.

Marj presented to Federal Express the number of letters received in and sent from the campus via Federal Express and the amount *not* delivered to the addressee by 10:30 A.M. (or refund granted, as their policy states) and Federal Express was more than willing to change their delivery route! They agreed to provide an earlier delivery time so that NCR Mailing Services could get the letters to the final addressee in the morning rather than Federal Express themselves trying to deliver each letter throughout the campus. Federal Express was asked to deliver by 9:30 A.M. each morning. They are now delivering by 8:45 A.M. Without any initiative from NCR, other overnight couriers have also been delivering earlier in the day, trying to compete with Federal Express's delivery service....

IMPROVEMENT

... As the Mailing Services QIT was looking for ways to improve the process, they examined the work activity to determine where they spent large amounts of their time. It was discovered that the manifest from the delivered Federal Express packages had to be written manually by the driver and the Mailing Services recipient, air bill number by air bill number, to assure that each line item was received.

Discussions with Federal Express unveiled that they were beginning to use bar coding to sort packages and that computers were being installed in the vans for inventory control. Suggestions and negotiations led the way to Federal Express using their bar coding system to preprint the manifest. Since the manifest is now being printed before the delivery occurs, the Mailing Services recipient can sign once for the entire delivery; it releases the driver faster and causes less interruption in Mailing Services. This simple change saves both NCR and Federal Express approximately 2.5 hours per week.

NCR Corporation has development and marketing organizations throughout the United States and around the world. The corporate mail service process consolidates and sends daily mail to each location from employees throughout the Dayton campus. The USPS Priority Mail Service guarantees deliveries within three business days. Mail sent on Mondays should have arrived prior to Thursdays but it didn't. The investigation started out as problem resolution and promptly turned into a process improvement. NCR's mailing envelopes are plain, light gray in color with the NCR logo and address printed only on the address label. Working with the USPS, Mailing Services learned that these envelopes with priority mail stickers front and back were probably getting mixed with the third class envelopes. The USPS had just developed a new envelope for this purpose with bold red, white, and blue banners and with "Priority Mail" printed on it so that it would not get mixed with other mail.

These envelopes work as planned but they were also very expensive. Marj contacted a manager at the USPS headquarters and explained the situation.... The next day the envelopes were delivered free of charge.... The process improvement is saving approximately $27,000 per year because NCR does not have to buy the priority mail envelopes from the USPS or NCR gray envelopes. From an effectiveness view, the consolidated mail process has severed one to three days in transit time through the U.S. Postal Service.

It is important to explain that these savings were gained during a period when the overall volume of mail increased by approximately 500 pieces of mail handled each day.

In addition, other savings resulted from these improvements: a $72,000 per year avoidance in overtime, a replacement of two full-time employees with part-time help, an 18 percent increase in sorting accuracy and improved internal delivery both on-site and off-site.

The National Cash Register Corporation has implemented the five-dimension approach on a broad scale throughout the company both in its domestic and international facilities. It has applied the model in various operations ranging from administration to manufacturing and product development and has achieved significant improvements in process effectiveness and efficiency.

AT&T's four-stage model

A four-stage process management model was developed at AT&T as part of their process quality management and improvement (PQMI) methodology for implementing the quality system designed to support the company's quality policy.[9] The AT&T approach was derived from IBM's process quality focus work. The four stages are (1) ownership, (2) assessment, (3) opportunity selection, and (4) improvement. These stages encompass many of the basic steps of process management.

1. *Ownership.* The ownership stage ensures that a process owner has been identified—a person accountable for the process who can manage it across organization or function boundaries. This stage also involves the establishment of a process team. Team members represent the owners of the subprocesses comprising the process selected. In addition, team roles and responsibilities are defined in this phase.

2. *Assessment.* The second stage is composed of three steps. The first step is defining the process and identifying its output in terms of customer requirements. The second step involves determining and establishing process measurements. Step three assesses conformance to requirements. This stage provides an understanding of how the process functions, develops internal and external customer requirements, and defines activities within the subprocesses that are to be measured and controlled to meet these requirements.

3. *Opportunity selection.* In this step, an investigation is performed to identify potential improvements. Process problems that affect both customer satisfaction and product or service cost as well as simplification opportunities are identified and prioritized based on

customer satisfaction and business objectives. Improvement projects to be pursued are identified by process teams.

4. *Improvement.*   In this fourth and final step, a quality improvement team is organized and develops an action plan to address the improvement projects identified in the previous stage. Root causes of problems are isolated and removed and the improved process is monitored and assessed on an ongoing basis.

These four stages are embodied in the following seven steps of PQMI, which are intended to be used as guidelines by the process teams:

1. Establish process management responsibilities.

2. Define the process and customer requirements.

3. Define and identify or establish measurements reflecting requirements.

4. Assess, through measurements, conformance to requirements.

5. Investigate and identify process improvement opportunities.

6. Prioritize improvement opportunities and set objectives and criteria.

7. Improve process quality.

Based on course modules developed at its Bell Laboratories Kelly Education and Training Center in New Jersey, the PQMI methodology has been taught extensively throughout AT&T to thousands of managers and professionals and has been deployed throughout its organization. PQMI is consistent with the process management approach described in previous chapters. Steps 5 and 6, however, contain features of Juran's project methodology for improvement.*

Results accruing from the application of the PQMI approach have been significant. Improvements in customer satisfaction have ranged up to 90 percent; quality cost savings of six improvement projects totaled $5 million with an investment of only $500,000 (see ref. 10). Other, more recent process improvements are estimated to have saved AT&T tens of millions of dollars.[11] The following excerpt relates other important results and benefits of the process approach:

> The Business Process Management Teams brought forth numerous improvements in their products and services, e.g., sizable reductions in the

---

*The Juran approach is described in a number of books and articles. See, for example, J. M. Juran, *Managerial Breakthrough*, McGraw-Hill, 1964.

number of days required to process an employee reimbursement voucher, introduction of much more user friendly input forms, elimination of non-value added reports, etc. The middle managers are feeling more comfortable with their new team allegiances and skepticism about using Quality concepts and tools is fading. The reality of treating the Business Units as real customers is clearly taking hold. The PQMI methodology and competitive Benchmarking are widely viewed as helpful management tools.

The Business Process Management Teams are starting to get their front line supervisors and employees involved. They are doing so with a much deeper understanding of what the quality process is really all about. They're not "delegating" quality, but rather encouraging and supporting their people's involvement. Quality training is now being requested and Quality Consultants are viewed as very valuable facilitators. Slowly, but on a very discernable basis, attitudes of middle management are changing.

As a result of these efforts in the Controller organization, other major segments of the Chief Financial Officer's organization, namely, the decentralized financial groups supporting the Business Units and Divisions, began to adopt a Total Quality System utilizing the same key elements and strategies. This has led to the emergence of a total CFO wide view of customer/supplier interdependencies among all financial processes. Such a picture has provided a powerful tool for top and middle management in understanding their businesses, their inter-relationships and the customers of the processes...

Process Management is an excellent vehicle for facilitating change in management style and behavior by emphasizing the focus on the customer through a cross-functional, team based, preventive approach. It provides an ongoing structure and framework through which process improvements, control and planning may be orchestrated. Lastly, it can be done at the "strategic" level through empowerment and support of middle managers.[12]

The NCR five-dimension and AT&T four-stage models are similar in that both have ownership and improvement in common. AT&T's assessment stage contains the definition, measurement, and control dimensions of NCR. In AT&T's model, opportunity selection is a distinct stage whereas, in the NCR approach, selection takes place during the activities of the quality management team prior to definition. For all practical purposes, the two models are virtually identical except in method and are consistent with the basic process management paradigm of this book.

In the second half of this book, applications of the paradigm to a broad spectrum of business processes will be examined. The first chapter is devoted to cases in administrative and service functions followed by cases in financial operations and product development. Following these cases, various factors of designing a process will be examined. Finally, aspects concerning future ways of managing operations and organizations will be discussed.

Notes

1. R. F. Boedecker, *Eleven Conditions for Excellence: The IBM Total Quality Improvement Process,* American Institute of Management, 1989, Boston, MA.
2. R. F. Boedecker, "TQI—Conditions for Excellence," Quality, December 1991.
3. M. E. McGill, *American Business and the Quick Fix,* Henry Holt, New York, 1991.
4. "The Payoff from Teamwork," Business Week, July 10, 1987, p. 57.
5. "Motorola Revisited," Quality, May 1987.
6. E. H. Melan, "Process Management: A Unifying Framework for Improvement," National Productivity Review, Vol 8, No. 4, Autumn 1989.
7. M. E. Porter, "Competitive Strategy," Free Press, New York, 1980.
8. L. B. Albitz, "State of the Art Processes in Administration," Proceedings, IMPRO 89, Juran Institute, Wilton, CT. Excerpted with permission of the National Cash Register Corporation.
9. AT&T Quality Steering Committee, "Process Quality Management and Improvement Guidelines," Issue 1-1, Select Code 500-049, Publication Center, AT&T Bell Laboratories, Holmdel, New Jersey.
10. G. T. Shaw, "Process Quality Focus," Presented at the ASQC Business Process Improvement Symposium, Washington, DC, March 21, 1988.
11. G. T. Shaw, unpublished communication to the author.
12. J. W. Zachman, "Developing and Executing Business Strategies using Process Quality Management," Proceedings, IMPRO90, Juran Institute, Wilton, CT.

Cases in Process Management Applications

11

Process Management in Staff/Service Operations

In Chapter 9, we analyzed a frequently encountered subprocess in a staff department—document distribution—in terms of the seven basic steps of process management. We will now examine other applications of process management in product and service operations. In this chapter, we will describe three rather diverse cases in staff/service operations: product ordering, an administrative or staff-type activity called labor claiming, and a service operation, package delivery. Although each has some unique features, one can readily deduce common elements among the three.

Case 1: An Ordering Process: A Case of Late Delivery

One of the most critical components of a marketing operation is ordering—the result of sales activity. Ordering involves an interface with the external customer and with the internal operations of the firm. This case involves a large, multinational electronics company that provides business equipment to various industries. The equipment consists of several individual electronic units that are connected by means of electrical cables to form a system. A wide variety of cables are offered by the manufacturer because the systems are configured differently, depending on the type of equipment, the application, and the equipment layout. Cables differ in their length, complexity (number of wires), type of wires used, and type of connectors at each end of the cable. Hundreds of different cable part numbers exist. Each part number can be fabricated in a wide range of physical lengths. Tens of thousands of part number/length combinations can occur. Each cable is essentially built to order.

A customer can either be a first-time purchaser of the equipment or an existing customer who wants to reconfigure an existing system. In either case, interconnecting cables are required. These are ordered and billed separately from the electronic units, which are the main source of revenue for the firm.

As the company grew over the years, cable ordering progressed from a simple, manual ordering-information system using telex and punched cards to a more highly automated one called CABEX using display terminals in the sales offices.* About six months after the installation of CABEX, complaints started to come to the plant providing the cables from the sales offices. Most of these complaints centered around orders that were shipped late. Executive management began to realize that sales and revenue could be hurt if the plant continued to provide cables to customers in the present manner, because customer dissatisfaction was increasing. As a result, a senior staff person from headquarters knowledgeable in examining business processes was assigned to analyze and evaluate the situation.

This individual interviewed people involved in various aspects of ordering: management and personnel in the order department at the plant where cable orders were received, field personnel at several branch offices who made up and placed orders into the CABEX system, vendors who manufactured and shipped the cables according to orders placed by the order department, engineers responsible for the design of the cables, and warehouse personnel responsible for filling orders from stock. Based on these interviews, the following picture emerged.

Sales office personnel responsible for entering orders into the new system had not received any prior training on the functioning of the CABEX system and how to place an order. To many, the technical terminology and information formats appearing on the display screens of the computer terminal were new, unfamiliar, and unrelated to the ordering system of the electronic units. At the time of conversion to the new system, instructional documentation was not available to the order entry clerks. As a result, a rising number of calls from sales offices asking for basic ordering information began to occur. Within three months from the inception of the new system, virtually all personnel

---

*CABEX represented at the time a "re-engineered" solution to an existing process using the latest in network information technology. In principle, it represented a breakthrough improvement in capacity and cycle time over the existing, antiquated cable ordering process. As you will see, however, CABEX became a complex solution to a relatively simple ordering process.

in the order department of the plant—managers and employees—were spending most of their time answering inquiries from various sales offices around the country.

As the field interviews progressed, it was discovered that the manner in which cable orders were generated had also changed. Prior to the inception of the CABEX system, cable orders were developed by an installation engineer—a systems engineer who was resident in the sales offices and who either participated in or was responsible for the design of the entire system installation. This engineer had received training in the technical properties of cables that governed their ordering features. It was therefore customary for the engineer to prepare an order on paper for an installation and then turn it over to an order entry clerk for keying into the system. An analysis of ordering errors later showed that orders prepared by these engineers were virtually error-free.

At about the same time CABEX was instituted, the marketing function of the company had been reorganized and split into two divisions, one for marketing and sales of small and intermediate electronic products and one for larger systems. The former division was not staffed with these engineers as management did not feel that product complexity required these skills. As a result, cable orders for installations of smaller electronics systems were submitted directly by the sales person to the order entry clerk. It turned out that many orders of this type were replete with errors.

Interviews with field personnel also surfaced another change to the method of placing a cable order: sales representatives associated with larger accounts were bringing cable orders made up by systems personnel employed by the customer, thus bypassing the installation engineer. As a result, the accuracy of the order input varied, depending on whether the installation engineer or the sales representative provided the order to the entry clerk. Order quality was further influenced by the skill level of the clerk inputting the order. Clerks who had become experienced with the new system were less likely to create errors and place calls for help than newer entry clerks.

The interviews then moved from the sales offices to the order department within the plant, a group of 10 employees. This department was responsible for receiving and analyzing orders transmitted from the central CABEX ordering information system, determining if the orders could be filled from inventory, and placing cable orders with vendors where orders could not be satisfied by inventory. Here, it was found that one person in the department was dedicated solely to correcting order errors. An errata or exception report was issued weekly summarizing the data in each order found to be in error. In a typical week, more than 200 orders were in error for reasons ranging from blatantly wrong cable lengths (such as 1,000 feet instead of 10 feet) to improper connector types.

In addition, many of the employees of the department were spending most of their working time expediting cable orders (non-value-add activity) rather than performing required duties, such as balancing orders among the vendors. Virtually every day, calls from sales offices were being received inquiring about overdue deliveries. In turn, these employees would call the vendors making the cables and inquire about order status. After determining where the specific order was in the cable assembly process, the employee would call the person in the sales office to report this status. Frequently, the sales office people would request an earlier ship date or would be told that their order would be shipped late. In either case, because of the interdependecy between cables and the system, the specter of delayed completion of a sale of the system would arise. Among the management and sales personnel of the marketing organization, late closure of a sale was considered almost as serious as a lost sale. Consequently, sales representatives would assert considerable pressure on the manufacturing organization to achieve a desired delivery date. As a result, the vendor was frequently told to reprioritize an order.

The ordering situation had become so severe with system equipment that, at the request of executives in manufacturing, an allocation and priority system had been implemented with the marketing organization to minimize direct sales pressure on manufacturing management and personnel from the field. Because cable ordering was done through a separate ordering system independent of the main product, however, no priority mechanism existed. By the ninth month, management in the plant was receiving direct and unrelenting pressure from the field to adhere to requested delivery dates. Meanwhile, the problem of late deliveries worsened. Status reports were requested by management on a weekly basis.

After the interviews of the sales offices and order department were completed, it became clear to the analyst that no one working in the area had any idea of how the total process operated. A few fragments of work-flow description existed either in the form of memos or paragraphs on scraps of paper, but no overall description was available. A macro diagram of the order flow was constructed (see Figure 11-1). An activity level diagram was then derived from the macro and work descriptions elicited during the interviews. Boundaries were set in terms of the overall domain of the process and key interfaces were noted.

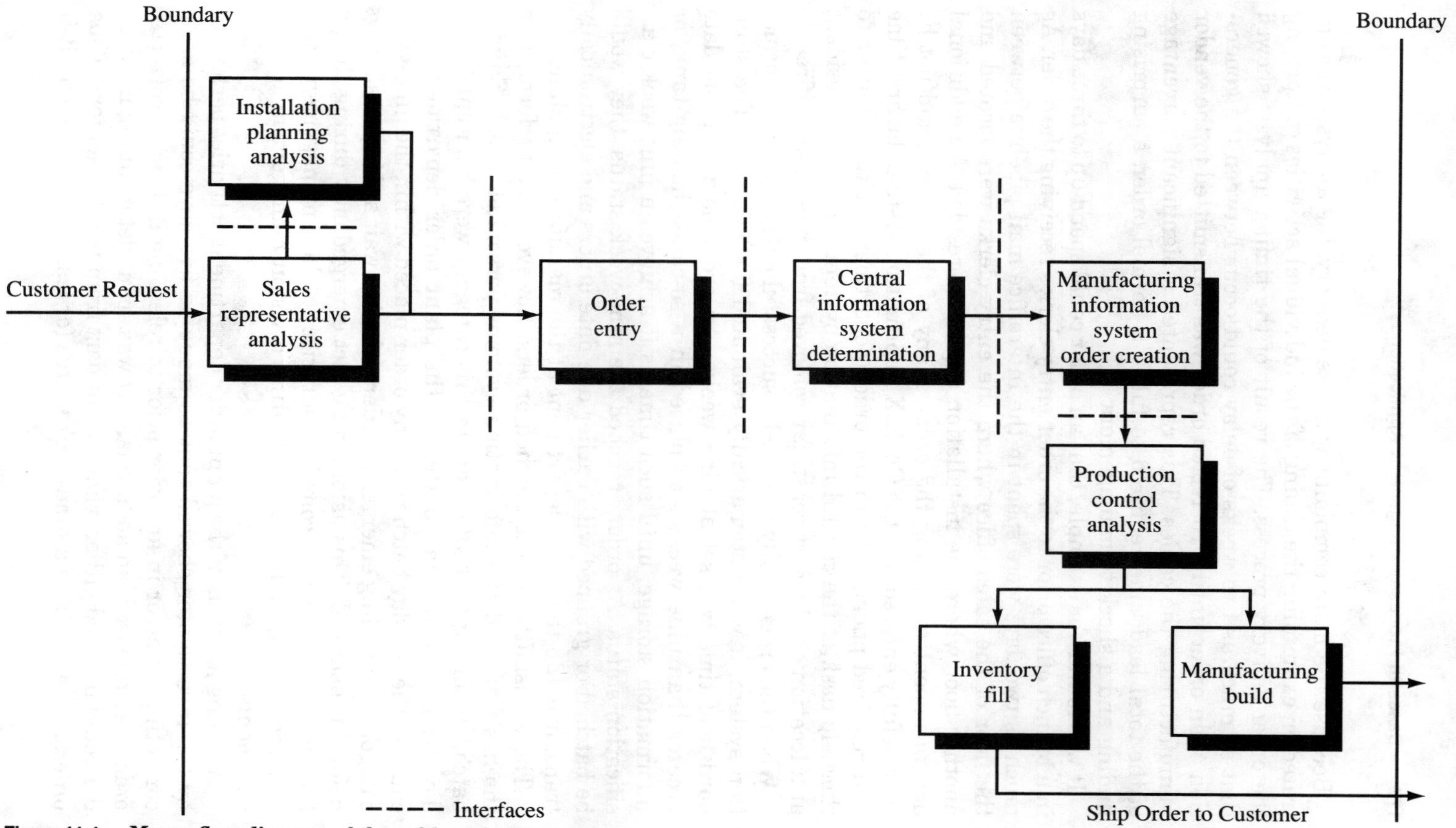

Figure 11-1. Macro flow diagram of the cable ordering process.

- - - - Interfaces

Because the main complaint was late delivery, the analysis concentrated on examining the timing of the individual activities comprising the order-delivery process. The result of the timing analysis showed that as much as two weeks of delay could occur between the generation of an order until a corrected order was transmitted to the vendor manufacturing the cables. This represented a significant percentage of the total lead time required to fulfill a normal order, a surprising finding and a shock to management.

It was found, for example, that an order could spend up to three days in a branch office prior to the order entry clerk's receiving the order. As much as two days were spent in the interoffice mail pipeline between the floor of the sales office where the entry clerks were located and another floor where the installation engineers resided. An additional day was spent in placing the order. Many of the orders could not be successfully entered into the CABEX information system the first time and required phone calls to the order department at the plant or to the help desk of the central information system department resident at a location on the east coast, far removed from most sales offices.

When the order finally achieved a successful entry into the information system, it was automatically evaluated for lead time. The standard lead time was set at nine weeks. An order having a due date greater than nine weeks was placed in a so-called holding tank, or information storage, until such time as it achieved a nine-week classification status. As orders reached the nine-week status, they would be batched or grouped with earlier due date orders and electronically transmitted to the order department at the manufacturing plant.

The transmittal of the batched orders, however, was performed on the basis of what the information systems personnel called "Sunday batch," meaning that all orders in this category were held until Sunday evening to be transmitted to the plant order department. Because of the Sunday batch rule, any order placed within the nine-week category—including emergency orders—had to wait as long as six days before transmittal. This astonishing fact emerged only during the interviews and came as a complete surprise to plant management, who were now being blamed by marketing division management for late shipments.

A timing analysis of the plant order department activities performed by the investigator showed that as many as five days could be spent correcting an error in an order prior to releasing it. Once a corrected order was released, an additional day was spent determining if the order could be filled either wholly or in part from plant inventory. Thus, orders, once reaching a nine-week status (or less), experienced built-in delays that ranged anywhere from an average of 5 working days for a defect-free order to a worst case of 12 days for orders containing errors.

The analysis then moved to process measurements. It was found that the main measurements used by management were the number of late deliveries, the number of orders filled, and the total number of cables made—all on a monthly basis. There were no statistics available on the number of orders placed that conformed to the required lead time and the number placed that violated the lead time. Similarly, there was no information available on the types of cables ordered or their physical length, so that a Pareto analysis on usage by cable type and length could not be performed.

Software programs were written to enable a statistical analysis to be performed on the existing order data base. The analysis showed the following:

1. As many as 85 percent of the orders placed each month did not conform to the lead time requirements published by the plant. On the average, in any one month about 50 percent of the orders violated the published lead time. This explained the extraordinary amount of expediting activity occurring in the plant order department. It was readily apparent that this process was operating in a reactive mode.

It also became clear from field interviews that sales personnel considered ordering cables to be an afterthought rather than an integral part of a system order. Hence, little attention was paid to published lead times.

2. A significant proportion of the late orders were revisions to orders previously placed, causing not only additional expediting but actual changes to work in process in the vendor's manufacturing facility. This, in turn, incurred associated modification and scrap costs. An examination of vendor production control records showed that as many as 30 percent of the orders on the floor were modifications to existing orders in process. Many of the orders appeared to be "dummy" or fictitious orders that appeared valid to the CABEX ordering information system.

Inquiries of sales office personnel showed that dummy orders were being placed at the time of the order for the main product mostly because of what was termed "system prompts." System prompts were instructions appearing on the main order entry system computer terminal screens. These prompts instructed the person entering the order to place an order for cables. In many cases, the system installation had not been configured, so the number, type, and length of cables could not be determined. The only recourse available to the sales representative to satisfy the ordering system requirements was to place a fictitious or dummy order knowing that the order would subsequently be changed. Later, as the system configuration was defined, an order change would be processed. Frequently, this occurred within the nine-week lead time rule and resulted in many instances of late deliveries and aborted orders.

3. An analysis was performed of the type and length of the cables in these orders to determine if some pattern existed. Results showed that nearly 65 percent of the orders involved only 15 percent of the total cable types available, another example of the Pareto effect. Within each type, a salesperson could order a specific length of cable, which then had to be custom-made to length. Technical consultations with cable application engineers indicated that, in all but a few applications, exact cable length was not a critical factor.

Hence, a proposal to make cables in specific length increments and stocking them for the most frequently used types was developed and implemented. This eventually contributed to a significant reduction in back orders and late deliveries.

4. An analysis was also performed of late orders over a 12-month period. Results showed that an average of about 53 percent of the orders were late because of defects in the actual order. Basically, the order entry input procedure lacked the information needed to process the order correctly. Further, the verification procedures in the order entry software were insufficient to prevent defective orders from being accepted by the host information system and subsequently transmitted to the plant order department where the defects were discovered.

As the investigation progressed, it was found that only installation engineers at the sales office (who were part of the field service division) received any training in choosing and ordering cables. Neither sales representatives nor order entry clerks received appropriate training.

Based on the analysis performed, the following actions were taken by the management of the function in which the plant ordering department resided. The management of this function, incidentally, had assumed ad hoc ownership of the order-delivery process.

1. A simplified, easy to use and understand cable-ordering manual was written and issued to all sales offices of the marketing divisions. Technical writers were assigned and instructed to work in conjunction with installation engineers to transform technical description into nontechnical terms using graphics and cartoons to explain how cables were to be ordered. After distribution of these manuals to the sales offices, numerous calls were received by the plant order department

complementing them on the manual's quality and ease of use. The number of inquiry calls began to decrease.

2. As a longer term action, the information systems group supporting the marketing division agreed to develop software for ordering cables that would be integrated into their current system's installation design program. This would eliminate all manual ordering for a new installation by generating an order electronically.

3. A lead time of 15 days was committed by the plant for orders where standard stock length cables were specified. These standard lengths would be satisfied by stocking them in inventory. It was anticipated that over 60 percent of the orders could be filled by standard-length stock. The shorter lead times were expected to induce sales personnel to use the non-custom-length option.

4. The plant proposed a so-called frozen zone strategy to executive management of the marketing division in order to reduce the number of changes to cable orders. In the frozen zone approach, changes would be honored by the order department for existing orders having a 20-day or greater lead time. Orders having less than 20 days lead time would be brought to a priority committee set by marketing management, who would decide which orders were to be prioritized. All orders with less than five days lead time (essentially emergency orders) would enter the absolute portion of the frozen zone where changes could be approved by executive management only.

The result of these various actions are shown in Figure 11-2. The average delivery time decreased significantly—from about 61 days at the start of the investigation to nearly 40 days within 12 months, a 33 percent improvement in response time. Order entry errors also decreased dramatically—over 80 percent of the expedite and inquiry phone calls were eliminated.

This case illustrates the ability to change and improve a large and complex process by means of process management analysis. This analysis, coupled with ownership initiative and management action, resulted in an impressive productivity improvement: for a relatively constant order work load, staffing levels were reduced by over 40 percent. The process became significantly more effective and efficient.

The case also shows how a large and complex cross-functional operation must be managed as a process to achieve effectiveness and efficiency. The key problems were a lack of understanding of how the process operated, its cycle time elements, a lack of understanding of the technical parameters required for ordering, suboptimization within the marketing function by circumventing ordering ground

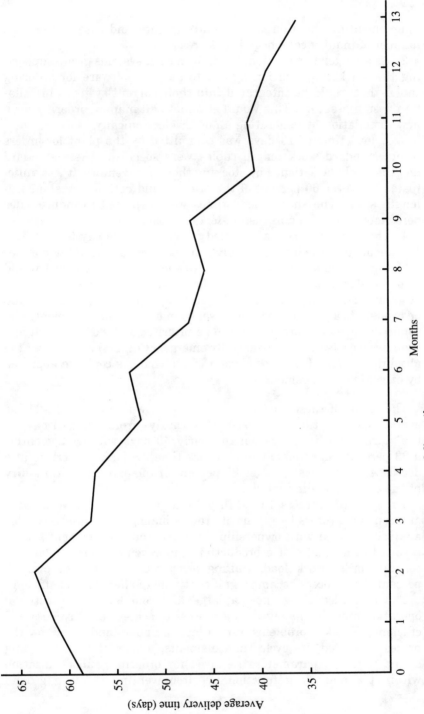

Figure 11-2. Results of process improvement on delivery time.

rules and lead times, and the complex, little understood operational features of an information system that compounded the delay. It is also an object lesson in what can go wrong with a re-engineered process that is introduced to replace an existing one.

Case 2: A Management Reporting Process: A Case in Wasting Resources

In any staff, service, or manufacturing activity employing resources of various kinds, information regarding their utilization is generally needed in order to facilitate proper management of these resources. Examples of such information would be equipment utilization and various types of labor reports such as absence, overtime, and labor claiming of different activities or projects.

This case involves several technical service departments at a product development laboratory of a large business equipment manufacturer. These departments perform technical services concurrently for several design projects. For accurate labor accounting, work performed on each design project must be reported weekly. The personnel in these departments were required to fill out labor-claiming cards every Friday morning indicating the time spent that week on each project using information supplied by the accounting department on a project code sheet. For example, an engineer or technician might write in 20 hours for project A, 15 for B, and 5 for C for time spent performing a mechanical analysis of various product designs.

These inputs were then given to either an administrative assistant, secretary, or the department manager and entered on individual department labor-claiming sheets for submission to the accounting department, where the sheets from various departments were collected and sent to another location 25 miles away for keypunching of cards and subsequent entry into a computerized labor reporting system. Figure 11-3 shows a macro diagram of this subprocess.

Over a period of time, the accuracy of labor claiming deteriorated to the extent that precise budget expenditure reporting became impossible. As a result, the problem was highlighted to executive management at the location by the financial function and, as an outgrowth of executive concern, a process management analysis was performed.[*]

Preliminary interviews revealed that this subprocess had no owner. No one could be readily identified who had both the authority and

---

[*]This analysis was performed by Beverly DeMott while at IBM Corporation, Kingston, New York.

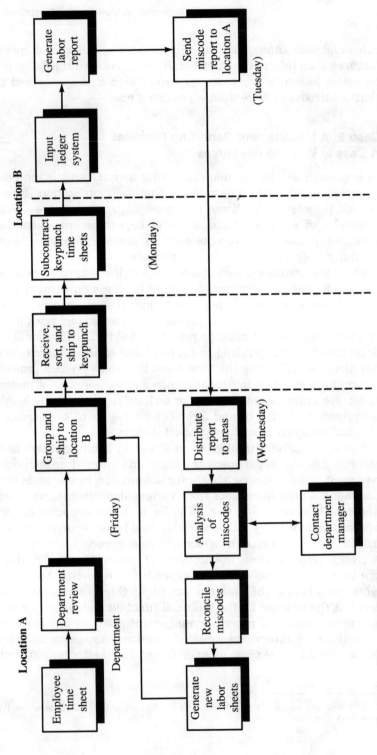

Figure 11-3. Macro of labor claiming process.

inclination to conduct corrective action. Further, there was no process description to show what was to be done in labor claiming. Finally, measurements were nonexistent. One could not determine the source or type of errors that were occurring. The operation was producing significant non-value-add time: a financial analyst was spending nearly all of his time between Wednesday afternoon and Friday morning on the telephone to department managers to correct errors and reconcile discrepancies appearing in the weekly labor claiming error report. Clearly, it was an unmanaged, ineffective, and inefficient process.

The flow diagram shown in Figure 11-3 was developed from two hours of interviews with people involved in labor claiming. It became clear that the only point of control that existed was the review of inputs at the time labor data was collected. However, this review varied from virtually no checking to a complete verification of the inputs, depending on time available, availability of a reviewer such as the supervisor, and the attitude toward quality work existing in the department.

It was also evident that no measurements of process performance existed. It was not possible, therefore, to determine the existence of a Pareto effect in the error distribution to see if certain departments or types of errors were predominant. In addition, root cause analysis based on quantitative data was impossible. Consequently, the first order of business was to determine a practical method for obtaining measurements.

A brief review of this subprocess showed that the output work product—the labor claim error report—could serve as a data base for analysis. Data was then extracted from the weekly error reports for the previous six months and summarized graphically in bar and trend charts for both defect category and departmental groups organized under second-level managers (see Figures 11-4 and 11-5). In all, several thousand data items were analyzed and categorized.

Errors were divided into three categories: source errors at the input, coding, and keypunching errors. Source errors involved using the wrong project number for input data. Coding errors were mistakes in both omission and commission, such as 200.0 hours instead of 20.0 hours claimed weekly against a project by an individual. Keypunching errors were mistakes in translating data on a source document to punched cards. Error data was also categorized by department and by groups of departments under a second-level manager. In all, nine departments were involved in this subprocess, consisting of over one hundred people.

Figure 11-6, a Pareto chart, shows that the majority of the input errors were attributable to a group of departments under manager Z. It also shows that nearly 80 percent of the errors were confined to groups managed by Z and M. From Figure 11-4, it is clear that

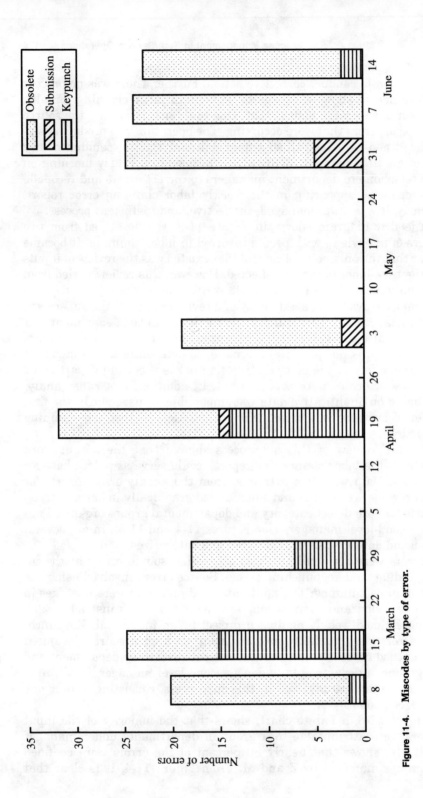

Figure 11-4. Miscodes by type of error.

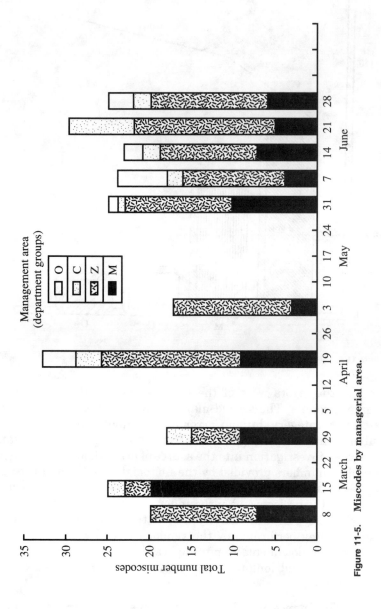

Figure 11-5.   Miscodes by managerial area.

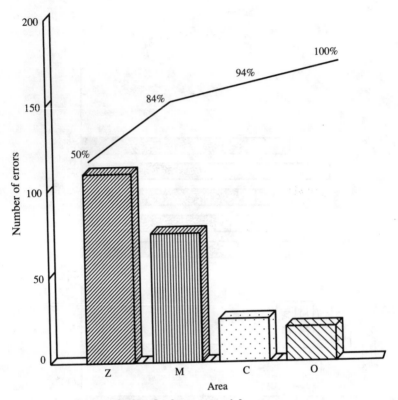

Figure 11-6.   Pareto of miscodes by managerial area.

most of the errors were of the type relating to the obsolete project charge accounts. The second major source of error was keypunching. The third category, human error at the input side of the operation (submission), was negligible—contrary to management expectation.

Further investigation into the source of error showed that the listing of project numbers provided by the financial function and used by the engineers and technicians was not up to date and did not accurately reflect all the projects currently under way in the laboratory. An analysis of the keypunching activity showed that punched card verification was not being performed by the vendor charged with the translation from written documents to punched cards.

The short- and long-term corrective action undertaken comprised the following:

1. a. The finance department responsible for the project number listings assumed ownership of this subprocess and began to issue revised lists on a monthly basis to all departments involved in charging labor to projects. This was an incremental improvement.

b. All departments adopted a common and more easily readable form sheet used for inputting data into the existing system—another incremental improvement.

2. Longer term, the electronic office information system currently in use in the laboratory was adapted to accommodate labor claim inputs and generation of labor charge lists by department. An electronic comparison of project charge number to be keyed in by the engineers and technicians against the current list of project numbers was developed and implemented. This created an instantaneous verification of input accuracy at the source. In effect, a point of control was established immediately at the input boundary of the process.

3. Utilizing the electronic office system network between locations A and B, subcontract keypunching was eliminated. Listings of labor charges were now transmitted each Friday directly to the labor reporting system at the company location issuing the report and bypassing a second source of error. Both this and the preceding actions were break-through improvements.

Results of the three major actions taken reduced labor claiming errors to zero. The revised process became more effective and efficient, productivity improved, and non-value-add time was eliminated. The payback ratio in terms of time savings was substantial:

Investment (hours)		Estimated hour savings (first year)	
Process analyst	4	Financial analyst	1200
Data analysis	96	Keypunching	400
Programming	80	Labor data gathering	
Miscellaneous	30	& transmittal	250
Total	210	Total	1850 Hrs.

Ratio (savings/investment) = 8.8
Note: the capital investment was zero

This operation achieved a defect-free state. Executive management now received a clear, accurate, and timely picture of project expenditures. Immediate action on overexpenditures could now be taken. No longer was the financial analyst required to spend as much as 16 hours a week investigating labor-claiming errors, nor did department managers have to waste time investigating and correcting these errors—all non-value-add activity.

This case is illustrative of many processes in which deterioration has occurred. This subprocess was essentially unattended. Error reconciliation was an accepted fact and nothing was being done to determine and correct the source of error. These errors resulted in a key part of the budget management process failing. This failure caused an inability to track project expenditures accurately. The process also lacked feedback on errors to people directly or indirectly responsible for them.

The process management analysis that was performed served as a catalyst for corrective action. Once the operation was bounded and defined, measurements taken, and data analyzed, root causes of the problem became clear. As the causes became known, both short-term corrective action and long-term breakthrough improvement solutions became evident. Interest by both employees and management in the problem made it easier to adopt short- and long-term solutions. After the initial review of this set of activities, detailed analysis work and corrective action was accomplished by individuals empowered to analyze and fix the problem. Other than to review progress and provide support to these people, managers played no directive role.

Case 3: Analysis of a Transportation Service Process: A Case in Managing the Interface*

The operation to be described is an example of a service process—the delivery and pickup of packages. For companies such as United Parcel Service and Federal Express, pickup and delivery are critical parts of their service system. This type of process is primarily a locational transformation—a coordinated and controlled movement of physical objects (packages) from one location (local terminal) to another through a network of intermediate storage areas known as distribution centers.

This case study involves freight delivery and pickup at a local terminal of a large, nationwide common carrier. The terminal receives the packages bound for its area (known as inbound freight), distributes the packages to customers (known as pickup and delivery or P&D), and picks up freight from local customers to be shipped to other parts of the country (outbound freight).

Because of the volume involved, the operation spans three shifts. It is recognized that shift-to-shift communication is extremely important in order to ensure proper coordination of both inbound and outbound freight and to assure proper and timely delivery. The terminal operates

---

*This case analysis was performed by Kevin Belsten, an MBA student at Marist College.

three shifts for each type of freight and is open 24 hours a day, six days a week. The inbound shift works from 12:01 A.M. to 8:30 A.M. and the pickup and delivery shift works from 8:00 A.M. to 5:00 P.M. The outbound freight shift works from 6:00 P.M. to 2:30 A.M.

Each shift has a supervisor whose responsibility is to manage the shift to which he or she is assigned. Combined, all activities comprise what is known as terminal operations. The terminal manager is responsible for the operation of the terminal as a whole. The shift supervisors report to the terminal manager. The employees are members of a national union. Inbound, P&D, and outbound operations can be considered as subprocesses of the terminal process. It is the responsibility of the terminal manager, the process owner, to ensure proper flow of all work that occurs within the terminal.

Each shift has a turnover period during which supervisors need to communicate with each other. During this turnover period, relevant information about the work is discussed and the operation plans are described to the incoming supervisor. At the start of the day, the inbound supervisor turns over operational responsibility to the P&D supervisor at 7:00 A.M. At the end of the day, the P&D supervisor will turn the operation over to the outbound supervisor at 4:00 P.M. Finally, the outbound supervisor turns the operation over to the inbound supervisor at 11:00 P.M., thus completing the cycle for the 24-hour period.

Recently, customer complaints have increased at this terminal. Most of the complaints concerned inbound freight not being delivered in a timely manner. A key indicator of delivery quality is a measurement known as "inbound local" (described later), which has been steadily rising at the terminal. The terminal manager wants these complaints eliminated.

A process management analysis was performed. Since ownership, boundaries, and interfaces were already established, the first order of business was to define the process.

Process definition

The activity sets comprising the operation are shown in the flow chart of Figure 11-7. The process is divided into three major groups of activities or subprocesses: inbound, P & D, and outbound. Each shift supervisor is responsible for ensuring that each activity is completed and for preparing for turnover at the end of the shift.

Inbound activities.   The first activity set, shown as I1 in Figure 11-7, involves receiving the inbound bills that designate each item of freight coming into the terminal for local delivery. These bills are inspected

Subprocesses

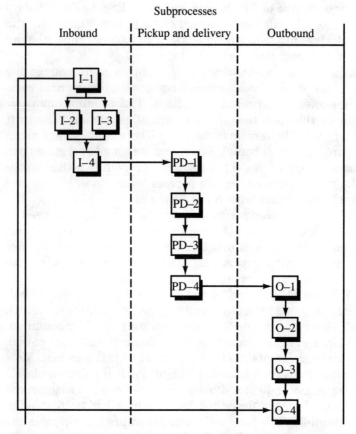

Figure 11-7.   Macro or activity set flowchart of terminal operations
process.

and delivery routes are designated according to the amount of freight
going to a delivery area. During I1, most of the planning is done for
the rest of the shift. In the second activity set, I2, the bills are coded
for each package. Each door of the terminal is assigned a number and
a delivery trailer is assigned to a door for each delivery route. Each
trailer is separated into four sections that are assigned numbers to
allow the loaders to determine the place in the trailer in which the
freight belongs. The code will include a door number of the delivery
trailer and a section of that trailer in which the freight should be
placed.

The code on each bill enables the workers to load a delivery trailer
in such a manner that the driver will be able to deliver the packages
without moving other freight as he progresses on his route. Thus, the

first delivery is on the rear of the trailer, the second delivery is just ahead of the first, and so on through the whole delivery route. The inbound supervisor must plan, organize, and direct the placement of freight in each delivery trailer for the package deliveries to take place efficiently.

In parallel with loading, any exceptions such as package overages, shortages, and damaged packages are identified and noted. This activity (I3) is performed during the night as a control function. Exceptions pertaining to any freight received are recorded on both the bill and on an inbound trailer manifest.

The final inbound activity, I4, is essentially a turnover period during which the inbound supervisor meets with the P&D supervisor, explains the delivery routes that are set up in each delivery trailer, and provides information about the delivery plans. If any changes need to be made, they are made at this time, before the drivers arrive to deliver the freight. I4 is basically an interface activity in which the status of the inbound work is communicated to the internal customer, the P&D supervisor. This is called a shift turnover in which management responsibility is shifted as well.

P&D activities.   Activity PD1 is performed by the P&D supervisor concurrently with activity I4. At this time, the supervisor takes over shift responsibility from the inbound supervisor and makes any necessary changes to the activities in progress. Delivery routes and package positions in the trailers are reviewed and drivers assigned to each route. The driver who is most knowledgeable of the area is usually assigned to that route. In addition, a driver's copy of a freight bill is separated from the terminal copy. The driver's copy is used for the delivery and the terminal copy allows the P&D supervisor to manage the delivery.

The next activity set, PD2, involves directing and controlling the delivery of the packages during the day. Each driver will communicate with the P&D supervisor (via two-way radio) the deliveries that have been made. Any problems encountered during this shift (such as shortages, overages, and damage) are noted by the supervisor in a log and compared with the inbound exception notes. The P&D supervisor will instruct each driver as to the appropriate route to take. Ideally, all the packages in the trailer are delivered the first time out. When this is not possible, the dispatcher decides which packages will be delivered and which will have to wait until the next day.

The third set, PD3, involves directing the pickup of packages (outbound freight). The P&D supervisor receives phone requests from customers during the day and assigns a driver to pick up the packages. The decision is based upon driver location and how much inbound freight has to be delivered. Ideally, inbound freight is delivered during the morning and early afternoon. The driver is then able to pick up outbound freight in the afternoon before closing time. Deciding the packages to be picked up and the timing of the pickup is the responsibility of the P&D supervisor. The service objective of the terminal is to ensure that all freight is picked up and delivered "on time and intact" to assure customer satisfaction.

PD4, the final set, is the turnover period during which the P&D supervisor informs the outbound supervisor of the remaining packages that need to be delivered and what, if any, outbound packages still need to be picked up. Information such as the amount of outbound freight and how much more is due is reviewed. A strategy is devised to consolidate the outbound packages for shipment to the so-called breakbulk terminals. Breakbulk terminals are distribution centers where freight is distributed according to geographic destination. The resulting work product of the P&D activities are delivered and picked-up packages.

Outbound activities.   All the outbound activity sets are noted by O1 through O4 in Figure 11-7. The first, O1, is performed by the outbound supervisor concurrently with activity PD4. The turnover from the P&D supervisor allows the outbound supervisor to develop a strategy to consolidate the outbound freight into as few trailers as possible for shipment to the breakbulk terminals. The supervisor also determines which trailers are to be loaded and assigns workers to load the outbound freight at this time.

O2 consists of loading the outbound trailers. Supervisors must ensure that the packages are handled and packed in an optimum fashion so that overages, shortages, and damage are kept to a minimum. Exceptions are noted on the outbound freight and recorded in a log and corrected before the freight is forwarded. During this activity, undelivered inbound packages are separated from the outbound packages and corresponding bills of lading are organized for submission to the inbound supervisor at the end of the shift.

During activity set O3, a determination is made of the inbound freight expected to be received at the local terminal. This is done by contacting a supervisor at the breakbulk terminal. It is important for the outbound supervisor to know the amount of freight that is coming into the local terminal in order to plan work for this shift.

Activity set O4 represents the final turnover period of the 24-hour cycle. At this point, the outbound supervisor informs the inbound supervisor what items are undelivered and what freight is due to come in from the breakbulk terminal. All the outbound freight has been separated and loaded into outbound trailers by this time. The supervisor inspects the inbound freight bills and begins to devise a strategy for the next day's deliveries. The inbound, P&D, and outbound activities are repeated in successive 24-hour periods. This process is fairly simple and self-contained and represents the predominant part of the terminal operation.

Performance measurements

The main process measure used by management is delivery performance, which is done daily. Delivery performance is reflected by a statistic known as "inbound local," which is a ratio of two numbers: a numerator consisting of a sum of the total number of inbound freight bills (*IF*), the number of "did not attempts" (*DNA*) (undelivered packages due to customer request or company convenience), and the number of undelivered items (*UB*). The numerator is divided by the total number of inbound freight bills. Inbound local, *IL*, is expressed by the following equation:

$$IL = \frac{IF + DNA + UB}{IF}$$

An *IL* value of 1.0 or 100 percent indicates that all the freight that came into the local terminal was delivered on the same day it arrived. Inbound as well as P&D activities have the greatest effect on this statistic. If the delivery plan is executed properly, *DNA* and *UB* will be low and *IL* will be close to 1.0. However, if the delivery plan is poor or is not carried out properly, this statistic will increase.

It was found that the *IL* statistic correlated with customer complaint calls. When the freight is delivered in a timely manner, the number of customer complaints is low. When delivery is delinquent, the number of customer complaints increases. Inbound local, then, is a surrogate measure of customer satisfaction.

Analysis and findings

Turnover activities (noted as I4, PD4, and O4 in Figure 11-7) from one shift to another involve information transfer by means of verbal communication. During the meeting, each supervisor requires the undivided attention of the other. It is essential for each supervisor to communicate his plans, expectations, and needs to the next shift supervisor. The terminal is continually busy because it is open 24 hours a day and is particularly hectic early in the morning and in the late afternoon. Adequate communication at these times is difficult, at best.

Investigation showed that a basic interface problem existed. The hectic environment at morning turnover and the need to transfer priority information on package delivery presented a weakness in this process. For service quality, it is imperative that the inbound supervisor know what packages have priority in order to expedite delivery. The P&D supervisor possesses this information. However, this information was not being transferred completely due to the informal (verbal) nature of the turnover and the interruptions existing at turnover time.

Corrective action and process improvement

A formal mechanism was needed to ensure that the information is accurately and completely transferred between the two shift supervisors. Upon completion of the analysis, two recommendations were made to the terminal manager: (1) implement a priority book (known as a "Hot" book) to record important information during the shift; and (2) provide an uninterrupted turnover period to allow a complete transfer of information.

Implementation was simple. The book provided written documentation for traceability purposes and acted as a reference guide for communication. An uninterrupted period for turnover would create a better condition for interpersonal communication. The supervisors would have the undivided attention of each other and an effective shift plan could be formulated without distractions. The P&D supervisor would fill out any important notes and information about inbound freight while speaking with a customer. During a turnover meeting with the outbound supervisor, the notes in the priority book are used to ensure that the outbound supervisor is knowledgeable about what needs to be done the following day. In turn, the book serves as a reference for the outbound supervisor during the turnover meeting with the inbound supervisor to ensure that priority delivery information is available at the start of the day shift.

Results

The recommendations were implemented. The P&D supervisor found it relatively easy to record customers' requests while speaking on the phone with them. The outbound supervisor found the turnover simpler when he had written notes and a period of time without interruption to accept the information. The inbound supervisor was pleased with the priority book because he had documentation to determine what occurred during the day.

In two months, the inbound local statistic decreased from 1.30 to 1.09, a 17 percent improvement (see Figure 11-8). The most impor-

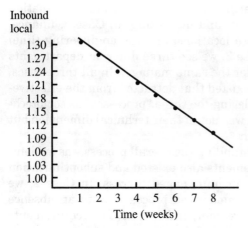

Figure 11-8.   Inbound local performance.

tant result of the new system was increased customer satisfaction. Although there was no formal manner of recording customer complaints, the P&D supervisor noted a marked decline in the number of complaint calls he received during his work shift. The improvement at the interfaces of the process allowed the inbound supervisor to create an improved delivery plan and enabled the terminal to progress toward its service goal of "On time and intact."

This case showed that the process was lacking in effectiveness because of interface problems. Customer complaints were increasing as reflected by weekly performance measurements. A critical activity was found in the process: turnover. Here, turnover involved mainly the communication of unfinished work at the end of one shift and the beginning of another. Addressing this resulted in an inexpensive solution to the problem and improved operational effectiveness.

Did efficiency improve? Considering the non-value-add time spent in answering customer inquiry calls and investigating the status of undelivered packages, efficiency did improve.

In this chapter, three diverse staff-service "processes" were examined from a process management point of view: an administrative process involving the placement of an order (Case 1); an administrative process involving labor claiming for budgetary reporting and control purposes (Case 2); and a service process involving scheduling and routing in the delivery and pickup of packages (Case 3).

These operations contain elements in common with other staff-service processes and elements that differ. The common element of the three is their cross-organizational characteristic. In Case 1, four organizational functions were involved: manufacturing, information systems, materials management, and marketing. In Case 2, finance, information systems at the two locations, and the engineering (line) function were involved. In Case 3, we see three different departments interacting in time (shift) under the same manager. In all three cases, work-flow interface problems existed that detracted from the effectiveness of these processes. Considering the three "processes" as sociotechnical subsystems, the problem was not in their technical dimension but in the social dimension of each.

In Case 1, a dearth of understanding the overall process was evident. A lack of knowledge of requirements also existed and suboptimization was clearly apparent. It was an unmanaged process. In Case 2, we see a similar situation of an unmanaged process due to an absence of ownership, definition, measurement, and control. The requirements for timely and accurate project code information were not communicated among the participants, nor was error feedback. Finally, in the last case, we again see that communication of operational information among the three shift supervisors was the key factor in the quality of the work flow.

In the next chapter, we will examine processes involved in the financial operations of a firm.

12

Process Management in Financial Operations

Introduction

Virtually every type of enterprise, profit and not-for-profit, has within its organization at least one function that contains some form of activity related to finances. Finance is considered one of the three basic functions of a firm; the other two are marketing and operations. In this chapter, we will examine how process management can be applied to typical financial activities encountered in the financial function.

Operations relating to monetary payment or funds outflow involve such core activities as general disbursements and payroll. These are sometimes organized into accounts payable. Funds inflow activities relate primarily to accounts receivable or monies owed by virtue of an organization's output of goods or services. Other activities such as cost accounting, cost estimating, and internal transactions (sometimes called general accounting) also comprise financial operations. In a typical firm, the financial function encompasses accounts receivable, accounts payable, general accounting, and treasury. In other organizations, activities such as cost estimating, financial control (auditing), and financial reporting may exist.

Transformations that are performed in this function are both transactional and informational in nature and the resulting output involves summary data and information of various kinds. Inputs involve both monetary instruments such as cash or checks and demand for payment such as bills and invoices. Integrity of financial information is a critical factor in financial operations. For example, comparing and verifying an invoice with a purchase order is critical to the integrity of an accounts payable process.

In general, accounts payable and accounts receivable are the most error-sensitive of all activities in a financial operation. For accounts payable, basic errors in payment can occur. Billing errors are quite common in accounts receivable and appear to increase in proportion to the complexity of the price schedule, the number of items involved, and, of course, the accuracy of the billing input data. In the cases described in this chapter, we will examine these two types of processes.

Figure 12-1 shows the typical activities involved in paying a bill in a firm—an accounts payable process. The process starts with a person in the requesting department of a firm filling out a form called generically a purchase request, which is sent to the purchasing function. The request is assigned to a buyer in this function. Depending on the nature of the request, the buyer may issue a purchase order directly for items involving no competitive bidding or may issue a "request for quotation" (RFQ) in which case some bidding activity is involved. The ultimate outcome of this request will be the issuance of a purchase order or orders to a supplier. A supplier data base or vendor file will also be updated showing this order.

Subsequent to the issuance of the purchase order, several sets of activities occur:

1. The supplier of the goods or service fulfills the request.

2. An internal document substantiating receipt of the goods or service is prepared and is used to update the supplier file.

3. An invoice or bill is sent by the supplier to the customer as a formal request for payment. Receiving, recording, and verifying the invoice represents an account to be paid. An invoice is a traditional demand for payment of services rendered or provision of some commodity. With today's information technology, invoiceless processing can be achieved, provided that means of verification of goods received and services performed are accomplished.

4. The final phase of the payables process comprises verification of the invoice, payment request and authorization, creation of the payment, and transmittal of the payment to the supplier.

The critical success factors of accounts payable are accuracy and timeliness. Accuracy is in terms of both payments for the billed material or service and validation of the charges. Timeliness refers to the cycle time between the receipt of the invoice and the transmittal of the payment. When accuracy and timeliness are poor, the process is ineffective and inefficient. This results in negative reaction from the customers of the process, namely, the vendors, as we see in the following case.

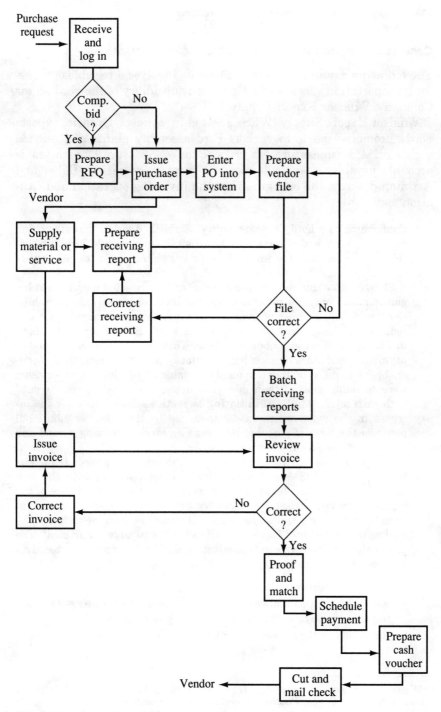

Figure 12-1. Accounts payable.

Case 1: Accounts Payable: A Case of Late Payments

The following example of process analysis involves a payables process for transportation services of a Canadian subsidiary of the Ford Motor Company, Windsor Export Supply. *

Windsor Export Supply (WES), located in Windsor, Ontario, exports Ford automotive parts to various Ford assembly plants and outside suppliers. A business process critical to WES's overall operation is accounts payable. A key subprocess is invoice processing. Baker and Artinian describe the background of the invoicing operation and situation as follows:

> Conforming to a Ford Company policy established for its manufacturing divisions, WES had eliminated its own audit function and hired an outside service to audit freight bills for proper charges prior to payment of carriers....
>
> Shortly after this change, the number of bills rejected for payment began to rise. Computer-generated queries had to be resolved before bills were paid. The number of queries increased to several pages; each one required a phone conversation or a review of microfiche records—both involving substantial amounts of time. This process frequently delayed payment (sometimes by months) to surface and air carriers; these carriers began to resist further shipments of material. In addition, supervisors were spending most of their own time determining "what went wrong" with each rejected bill and initiating corrective action. The workload increase in the department was compounded by the influx of additional paper in the form of past-due bills sent by carriers making second and third efforts to obtain payment....
>
> It was in this environment that an improvement project team was formed in May 1984 on the initiative of employees and with the support of management. Team members represented a diagonal slice of the organization and included supervisory and non-supervisory representatives from the WES traffic, parts control and accounting departments as well as the manager of Ford's Oakville accounts payable department. The group elected the traffic rate analyst as leader. A number of objectives were established, including:
>
> ■ Reduction in time to pay carriers.
> ■ Reduction in the number of phone conversations with carriers regarding overdue payments and the processing of past-due bills.
> ■ Reduction in time taken to audit and correct payments to carriers.

---

*This case has been excerpted from "The Case of Windsor Export Supply," E. M. Baker and H. I. Artinian, *Quality Progress*, June 1985, used by permission of ASQC.

At WES, an invoice proceeded from the carrier through the internal traffic department of WES to a subcontract verification service in New Jersey to a subcontract keypunch service at a New York bank to a San Francisco bank for compilation and finally to the accounts payable department at another Ford facility in Ontario (see Figure 12-2).

The authors describe how the invoice is routed and verified:

1. The carrier issues an invoice for services rendered.

2. WES traffic stamps received invoices and adds information that uniquely identifies the bill and allows the production of summary performance reports.

3. WES sends the stamped invoices to an audit service in New Jersey (subcontracted by the bank) for a 100% audit to ensure that the proper rate has been charged according to latest tariff structures.

4. Audit invoices are sent to the contracting bank in New York where key information is keypunched for transceival to the bank's computer located in San Francisco.

5. The bank in San Francisco compiles the invoice information for mechanized performance reporting. Each record is then transceived to the Ford computer in Oakville, Ontario, where the accounts payable activity is located.

6. Accounts payable verifies the invoices transceived by the bank to ensure that all key information is included and correct. If any information is missing or incorrect, the Ford mechanized system automatically rejects the record of the input and generates a query which is compiled in a report and sent to WES traffic (approximately two to three times per month) for resolution. If all the information is correct, the carrier is paid.

The team then examined the process in terms of the customer-producer-supplier relationship model (described in Chapter 3):

The process starts with the carrier sending a bill to WES traffic. The carrier can be viewed as the producer and WES traffic as the customer. As the producer, the carrier receives various inputs from its own internal systems. For example, "carrier resources to do the transforming" include accounting personnel, systems and data processing equipment and methods and billing procedures. "Carrier resources to be transformed" are actual, blank freight bills.

Value is added to the freight bill (throughput) by putting information on the blank bill such as date, shipment origin, shipment destination, cost per cwt., any applicable discount and net billing amount. The completed bill is the carrier system output.

The user (WES traffic department) has certain input requirements that must be met if the information it receives is to be useful. WES traffic must be able to uniquely identify the freight bill and to verify the authenticity

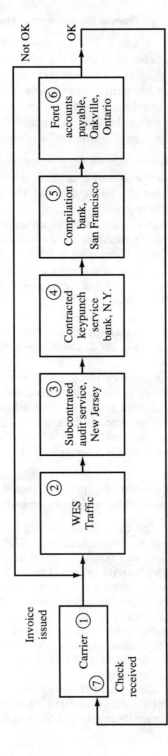

Figure 12-2. Invoice processing—before change. (From E. Baker and H. Artinian, "The Case of Windsor Export Supply," *Quality Progress.* Reprinted by permission of ASQC.)

and the amount of the charges. The carrier invoice issuing system should have output requirements related to the WES traffic (user) requirements. Each invoice should contain such specific pieces of information as description of material, quantity, weight, rate, special routing required and total charges.

Each set of producer-user interfaces can be analyzed in a similar manner. For example, the keypunch service (viewed as a producer) receives the audited invoices containing the information provided by WES traffic (e.g., carrier identification code, vendor code, traffic code, account distribution and circled amount). In turn, the keypunch service must consider the needs not only of the next user (San Francisco bank) but of subsequent users. It must know that Ford accounts payable will use the information to issue a check to the carrier for services rendered and to maintain financial control. To enable accounts payable to do this, the keypunch system must ensure that each completed record identifies each bill uniquely, contains every piece of information necessary for payment and is keypunched correctly.

The carrier is the final customer in this system. As with many administrative and service systems, the final customer and initial producer are the same.

The team next turned to performance measurements and feedback.

Quality performance requires an external feedback loop from the customer back to the producer.... In this system, the carriers (at stage 7) provided feedback to WES traffic department through telephone calls, letters, past-due bills, and personal visits. In almost all cases, the feedback was the same: The carrier expected prompt payment and that expectation was not being met. As a direct result, customers were growing dissatisfied....

The WES project team decided to establish a performance baseline by evaluating process control of the following operationally defined outcome: The number of days elapsing between the date an invoice was stamped as received by WES traffic (stage 2) and the date a check was issued by Oakville accounts payable (stage 6). This segment of the process was selected initially because the data were easily obtainable....

Initially, data were collected on batches of invoices that had been processed through WES traffic each month from September 1983 to June 1984. $\overline{X}$ and R charts were developed and showed the system was in statistical control but at a level that was too high.... It took an invoice an average of nearly 14 days to pass through this segment of the system. When that lag was combined with other factors such as the mail system, it turned out that many carriers were being required to wait 35 days or more for payment.

The project team developed an Ishikawa cause-and-effect diagram and identified several major reasons why bills were being rejected from the mechanized system.

Keypunch errors. During the actual keypunching, three types of errors tended to occur. The first type was an incorrect carrier identification code entered into the record. The code, consisting of four letters, was often entered in error when there were two different carriers having similar names (e.g., Brown Transport (BRNT) and Brown Express (BRWN)).

The second common type of keypunch error was an incorrect dollar amount entered into the record. On many occasions, the dollar value actually entered was the discount instead of the net total of the bill.

In the third type of keypunch error, the last two characters of the airbill number would be truncated and not entered into the system. The net result was that the airline would be unable to identify the specific bill Ford was paying.

Misfiling. Bills pulled from a "U.S. funds" batch were incorrectly replaced in a "Canadian funds" batch, resulting in payment in the wrong currency. This type of error occurred between the subcontractor and the bank.

Missing carrier/vendor codes. Invoices listing a carrier or vendor that was not on the master list in Oakville would be rejected for payment. This problem occurred if the master list had not been updated when a new carrier was added.

Lost or misplaced bills. Invoices were passed for payment by WES traffic but did not reach Oakville. Bills could be traced via In/Out logs at the audit firm but would subsequently disappear. In one instance, five batches of invoices returned from the bank had 400 missing bills worth approximately $185,000.

Ford management and the bank held meetings to resolve the problems. After several weeks, it became evident that progress was not being made and that prospects for meaningful improvement in the near future were not good. It was at this point that Ford management, on the recommendation of the project team, decided to bring the audit functions in-house and make several changes. WES management met with Oakville accounts payable personnel and, with the knowledge gained from the systems analysis, carried out a smooth transfer of the audit functions in August 1984. The new system ... incorporated the following changes:

- The audit function is now performed by the WES traffic department instead of an outside subcontractor.
- Previously, invoices passed for audit and sent to the bank were recorded to "fit" the bank's system. Now, invoices are microfiched immediately and then sent to keypunch with no intermediate encoding.
- The keypunch service is located across the street from the accounts payable department.
- There are now fewer stages through which the invoice information must pass. One of these stages, invoice microfiche, is simply a pass-through and usually takes less than a day.

Figure 12-3 shows the improved process. These changes resulted in a much simplified and more responsive operation: invoices were now encoded only once instead of twice, which reduced error and coding time; invoice transmittal distances were shortened, which substantially reduced in-process time; and the total cycle time shortened dramatically from 14 to 6 days. In addition, communication improved and, because of performing the audit function internally, increased control over the process was achieved.

This case illustrates several problems commonly encountered in non-manufacturing processes:

1. It is apparent that process ownership was lacking. There was no overall management and coordination of the activities. Changes had been made without regard to their impact on process effectiveness (i.e., customer requirements).

2. Activities such as auditing, keypunching, and compilation were separated by substantial distances from the input and output ends of the process as well as from each other. As a result, significant interface problems occurred. Internal customer-supplier relationships and requirements definitions appear to have been absent.

3. Overall process objectives (such as payment cycle time) were absent, as was process definition.

4. Process measurements did not exist. Rather, measurements were implemented to help define the problem.

5. The process operated in a reaction rather than a control mode.

Examining the operation by taking a process view—analyzing the entire process in terms of input-output and value-add transformation, looking at interfaces in terms of customer-supplier relationships, and measuring performance of key activities—provided the means for achieving significant improvement.

Case 2: Accounts Receivable: A Case of Timely and Accurate Billing

In this case, we examine another key process in a financial operation: accounts receivable. In large, multiple product or service organizations, accounts receivable presents a relatively complex set of activities crossing many functions. Accounts receivable generally consists of two primary subprocesses: billing and collecting. In billing, the charges to be claimed are computed and sent to the customer. In collection, payments for items billed are received, analyzed and accounted for,

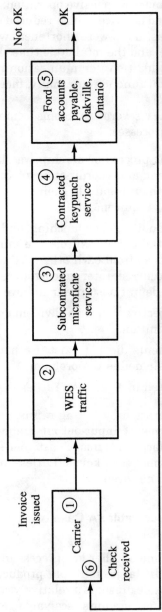

Figure 12-3. Invoice processing—after change. (From E. Baker and H. Artinian, "The Case of Windsor Export Supply," *Quality Progress*, June 1985. Reprinted by permission of ASQC.)

and funds deposited. In the case described, we see how this process is dissected for a service firm. The following is a process management analysis of its accounts receivable operation.[*]

Overview.   This firm performs engineering services of various kinds that are billed to clients. A detailed breakdown of services performed is provided to the billing department by engineering every week. This information is compiled monthly, at which time a draft invoice is produced and distributed to managers for review prior to billing the client. It is important for the image of the firm that management review all documents for accuracy prior to their being sent to the client. After reviewing the preliminary (draft) invoices, project managers return them to the billing department with any necessary adjustments. At this time, a final invoice is mailed to the client and the invoice amount, with project number, is given to the collections department for follow-up and processing of the receivables.

This process begins with billing information on services performed and ends with a paid invoice from the client. The owner of the process is the manager of accounts receivable.

Process definition.   Referring to Figure 12-4, the billing department provides information to three customers, two internal and one external: collections, engineering services, and clients. On a monthly basis, draft invoices are produced and distributed to project managers. Input for these invoices comes from two sources. The first is the weekly time sheets filled out by all employees. Time sheets are used by billing personnel to assign time spent on all projects. The other is a billing form sheet which is prepared by project managers for all projects at their inception. This form sheet provides details of exactly how the client will be billed. These two inputs are used by billing personnel to produce a preliminary, or draft, invoice that is identical to a final invoice. Draft invoices are then sent to project managers for review and adjustment prior to producing a final invoice, which is then sent to the client.

After the draft invoice is reviewed and any adjustments made, the final invoice is produced. At this time, any additional paperwork required by the client is prepared. This additional paperwork could be a letter, supporting documentation, a voucher, or a billing summary. Once all paperwork is complete, the invoice is mailed to the client.

At the time the final invoice is sent, the invoice amount and project number information are sent to the collections department. Collections

---

[*]This analysis was performed by David Damon, a graduate student at Marist College.

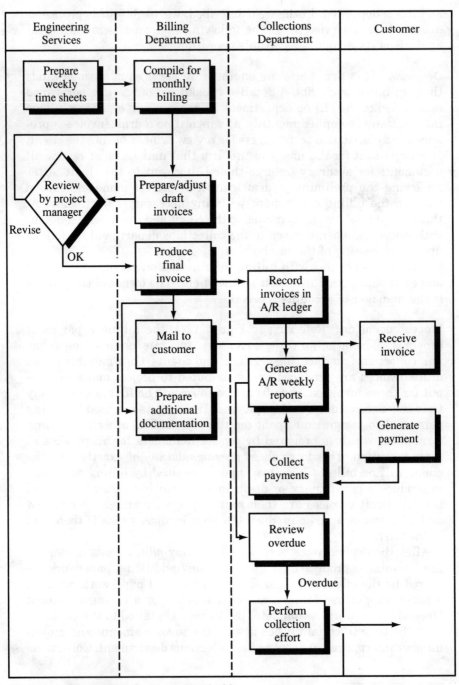

Figure 12-4.   Macro flow of accounts receivable.

is responsible for maintaining the accounts receivable ledger, receiving all checks mailed to the company on a daily basis, and depositing these receipts daily. It also codes these receipts against the invoices.

This department is also responsible for generating and distributing accounts receivable reports to management and for making any collection efforts necessary. Invoice project information as well as paid invoice data from daily receipts are used to produce the report, which is sent to project managers and collection personnel weekly. The managers and collection people then work together to correct any errors and collect any amounts that are overdue.

Points of control.   There are three points of control in this process. The first occurs at the time that project managers review draft invoices. At this time, a careful review of the charges listed on the draft invoice should uncover any charges that are not billable or have been charged to the wrong project number. The second occurs at the time that billing adjustments requested by project managers are checked for validity by billing personnel. For example, a project manager might try to keep his project from overrunning budget by transferring charges to another project. Checking with the project manager of the second project can avoid this problem. A third control point exists at the review of the accounts receivable reports. A review of this report by collections personnel and project managers generally uncovers any errors that may have been made in recording invoices in posting cash receipts to the accounts receivable reports.

Measurements.   Five basic measurements are used in this process. The first involves the time required of the project managers and billing personnel to produce a final invoice. A three-day turnaround by project managers and a one day turnaround by billing personnel is allowed. Failure to adhere to this schedule results in a delay in collection that affects cash flow. Measurements showed that, by and large, the turnaround targets were not being met, resulting in most invoices being sent late.

The second is an error measurement that involves comparing the amount on the draft invoices with that on the final invoice. At times, the final invoice amount was lower than on the draft. Failure to bill all charges tends to be an indicator of potential project problems. If these problems turn out to be major and items are not billed, there is a negative impact on cash flow and revenue of the firm.

The third is a ratio of the total number of draft invoices to the number adjusted. A high percentage of adjusted invoices indicates carelessness in recording time spent working on a client's project. Each adjusted invoice represents time wasted in correction. In a perfect process, all charges would be initially correct, eliminating the need for management review and billing adjustment. It was found that over fifty percent of the invoices required adjustment.

The fourth measurement is the aging of accounts receivable balances. Balances in the overdue section of this report can indicate problems with a project, a client, or a project manager. A large amount overdue could indicate that the money may never be collected, causing a loss of revenue to the firm.

The fifth process measurement is the average collection period, a frequently used measurement for receivables. This statistic calculates the average number of days required to collect receivables. In general, the lower the collection period number, the more efficient the process. It should be kept in mind, however, that an efficient collection process could result in an ineffective one; an aggressive collections policy might be perceived negatively by the customer.

Assessment.   In general, the analysis showed that the current process is inefficient. The process can be improved by examining time delays in each activity, because the response time of receivables is a critical internal success factor. The analysis showed three areas of process improvement that are required.

The first and most important area is the adjustments that are made to draft invoices. The current adjusted-invoice percentage is substantial and causes additional overhead labor in making these adjustments. The solution to this problem is to assure that all project employees provide error-free time sheets. A second improvement involves the turnaround time for transforming draft invoices into final invoices. While the allowed time is four days, very few invoices were actually produced within this time. The solution to this problem is to require project managers to review draft invoices the day after they are distributed.

The third area involves the time required to collect receivables. This time is currently excessive, mainly because of the slow turnaround time of invoicing. Some clients commented, "If it takes thirty days to bill me, I'll take ninety days to pay you. How can you require me to be responsive when you aren't?" Another reason is an unwillingness to create negative feelings in clients by demanding payment before it is overdue. Since the field is very competitive, it is unwise to lose a large client because of a payment issue. The solution involves shortening the billing cycle and making payment terms an important part of the contract with the customer.

The two cases described in this chapter show that financial operations can be analyzed by using the process management approach. In the first case, we saw how, by a team effort, the customer-producer-supplier relationship model can be applied to resolve input and output requirements in a large and complex accounts payable operation. This case example showed how interface problems seriously affected the response time and ability of the process to the extent that the vendors as customers of the process began to react negatively to being paid late.

In the second case, a relatively simple receivables process for a service firm was analyzed showing the close relationship between the billing and collections subprocesses. The overall performance of this process cannot be measured solely by efficiency and effectiveness indicators. The manner in which the collections subprocess is conducted is highly dependent upon how the organization relates to its clients and, in the final analysis, this relationship is governed by business judgment and the policy of the firm. Although the receivables process described is for a small service firm, much of it contains activities common to the receivables of any type of organization.

In the following chapter, we will examine product development in terms of a process. Here, again, we will see that process management can be applied successfully to analyze and improve operations not even remotely resembling operations in administration or service. Again, we will see symptoms common to any ineffective and inefficient process.

13

Process Management in a Laboratory

This chapter shows how process management can be applied to operations within a development laboratory in the design of products. Development generally consists of a series of activities that comprise the design and release of the product for manufacture. With today's complex processes, design is rarely a simple matter of developing drawings and specifications of the product to be made and sending them to the manufacturing organization. Designs are based on a technology or set of technologies that are deemed appropriate and feasible for an intended application. Requirements are defined, negotiated, and often renegotiated prior to (and even concurrent with) design and prototyping.

In recent years, managers in development organizations have begun to realize that the traditional barrier or "wall" between development and manufacturing needs to be eliminated. This barrier has been a prime example of suboptimization that has resulted in ineffective and inefficient designs being released to production.

The concept of concurrent or simultaneous engineering was developed to eliminate or greatly reduce this barrier. In this concept, all organizations involved in the manufacture and service of the product participate in its design and prototyping. Representatives of procurement, manufacturing, and various engineering functions such as manufacturing, industrial, quality, and equipment engineering are assigned to assist in the development of the product. Each of these representatives helps to develop a design from the viewpoint of their own organization. For example, a field service representative looks at a potential design from a serviceability or ease of repair point of view.

A manufacturing or manufacturing engineering person would look at the design from a producibility viewpoint. Each organization represented is responsible for specifying the design requirements reflecting its mission, purpose, and objectives. In a sense, they are the internal customers of product development.

Upon successful evaluation of prototypes, preproduction models may then be evaluated and analyzed in terms of the intended application prior to approval and release to manufacturing. In certain development organizations where checks and balances exist, an independent assurance group may evaluate and qualify the design for its intended use. A macro view of the product development process, including design qualification, is shown in Figure 13-1.

In cases where it is economically feasible (or strategically appropriate), these models may be field tested to evaluate performance under application conditions. In other cases, models fabricated in a manufacturing environment may be tested. Both approaches have the advantage of early discovery of design deficiencies. In cases where a design may be one of a kind—such as an atomic reactor or a power plant—the use of prototypes is not feasible, and heavy reliance is placed on mathematical modeling, simulation, standards, and design reviews to assess and qualify the design.

Development is inherently a trial and error (or recursive) process. When a design is subjected to evaluation, review, audit, or modeling, various kinds of deficiencies or oversights are often discovered. These deficiencies are investigated and cause(s) determined. The design is corrected, a new evaluation performed, and the cycle repeats until a satisfactory design is achieved, after which the development process continues to the next phase. In some cases, this trial and error process has become cost prohibitive and techniques such as design of experiments and the Taguchi method alleviate some of the disadvantages of the empirical approach.

Development Phases

The development process typically takes place in four distinct phases:[1]

1. *Requirements.* Requirements may be customer-based, may originate internal to the organization, or may be a combination of the two. Product requirements are generally defined prior to the prototyping phase but are not necessarily completed by the start of it.

   In recent years, the requirements portion of the development process has received much attention. The concept of quality function deployment (QFD) has been widely publicized in the literature[2,3]

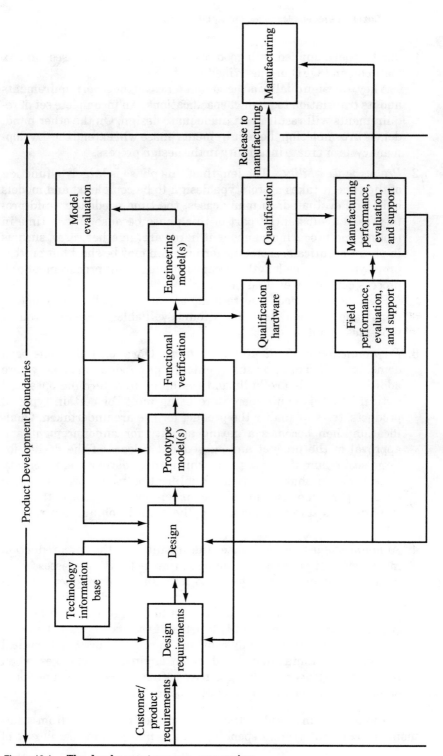

Figure 13-1. The development process—macroview.

and has been applied with good results in a number of cases. Matrix methods and QFD are described in Chapter 3.

A key to design quality is the completeness of the set of requirements and its translation to product specifications. An incomplete set of requirements will result in an inadequate design. On the other hand, delays in completing the requirements phase will extend the development cycle or create instability in the design process.

2. *Prototype-feasibility.* The length of this phase is largely a function of the time it takes a prototype design to be completed and models built and evaluated. In many cases, the time needed to build prototypes is a substantial part of this phase because of lead time in tooling and procuring component hardware. In other cases, such as in pharmaceuticals, prototype formulation may be short but process prototyping may be lengthy because of equipment procurement and process capability evaluations.

   Statistical methods and techniques such as evolutionary operations[4] and Taguchi[5] have been found valuable in creating robust processes and designs.

3. *Preproduction-evaluation-qualification.* Once feasibility has been demonstrated, a design usually enters a preproduction phase where additional models are built and evaluated to determine on a statistical basis if requirements are being met. For certain types of products, tests to qualify the product for use are undertaken. Qualification then becomes a gating activity for announcement and approval of the product and subsequent release of the design for large-scale manufacturing, at which time development activity phases out. In pharmaceuticals, qualification may represent a substantial portion of the entire development cycle because of the need to perform tests on humans and the time to obtain government approval.

4. *Announcement-release.* Once the results of the preproduction model evaluation and any qualification activity are assessed, a management decision is made to announce the product. At this time, manufacturing activity increases, production line scaling or buildup occurs, and product begins to emerge from the factory. Although development usually finishes at the end of phase 3, follow-on product engineering may continue in order to provide technical support to manufacturing and make engineering changes when required. Engineering changes may be made to correct functional or safety deficiencies or to reduce product cost.

The total time involved in these four phases may range from a few months to several years, depending on the nature and complexity of the product as well as business and technical factors. Each of the four phases may involve iterations or delays for various reasons. Of major interest in the design process is its timeliness, a critical success factor. In today's environment, the time in which a design is translated from product requirements to announcement and manufacture has become crucial from a competitive viewpoint. The speed with which new products enter the market is critical in the battle for market share.[6] Nevertheless, the various factors that affect the length of the development cycle are often not apparent to management. Process management provides an analytical tool to enable a developer to examine the factors affecting this cycle.

In the following case, we see how process management is applied to provide significant improvements in the development cycle of a major manufacturer.

Case 1: The Development Process: Cycle Time Analysis*

The case involves a large manufacturer of business equipment and the design of a critical subassembly of this equipment. Management in charge of its design and release encountered schedule uncertainties and requested an assessment of the development process to determine if delays would occur and whether improvements could be made in the design cycle.

A process management analysis and assessment was performed. The strategy employed was to use the team approach involving participants knowledgeable of and working in key parts of the process. The author played the roles of facilitator, coordinator, and team leader. Process ownership was established first. A second level manager was identified as the owner because various design activities converged at this management level. Establishing ownership at the outset of the analysis promoted management involvement and, ultimately, the adoption of improvement recommendations.

The next step was to identify process boundaries. In this case, the process started with the request to design the subassembly. The output boundary terminated with an approval to release the design to manufacturing by the qualification department, an independent group whose mission is to evaluate, analyze, and assure the design for customer use. The boundaries are shown in Figure 13-1.

---

*This case was based on the author's paper presented at the ASQC Quality Congress in Toronto, Canada, May 1989. It appears in *AQC Transactions* under the title "Achieving Quality Excellence in Development."

The third step comprised defining the process in terms of its various work activities. An activity level flowchart comprising some fifty key steps was completed. Typical of many processes, the design process in this instance was totally undefined. No one had a detailed view of either the subprocess components or the total development process. Process definition provided several immediate results.

First, the overall process with its subprocess relationships became visible to management. Second, all of the individual activity elements became known. Critical or gating activities were identified, such as qualification, which affected the overall process response time. Third, the flowchart defining the process became the medium of communication for those involved in the design. Fourth, the flowchart provided a medium for educating those new to the process. Finally, and most important, the very act of defining the process provided the participants with an insight into improving it. In fact, the analysis resulted in ten key improvements (described below) that reduced the overall design cycle and resulted in a more efficient and effective process.

The fourth and fifth steps involved establishing control points and measurements. Several control points were used as measurement points for determining process cycle time (measured from the time when customer design requirements for the product were agreed upon to the time at which qualification was completed) as well as assessing effectiveness and efficiency. In this case, measurements at control points such as design reviews and model evaluations, as well as various work completion events were used to measure response time.

The sixth and final step comprised corrective action and improvement. Once the process was defined and its response time measured, it became clear that corrective action was required: the overall design cycle exceeded planned development schedules by a factor of two. The analysis itself provided a baseline for improvement and, combined with assessment sessions with the process team, a final set of ten suggestions to improve the process were developed and presented to management.

Of the total of ten incremental and breakthrough recommendations developed by the team (all of which were approved by management), five involved basic changes to the design and qualification subprocesses, which served to reduce overall design cycle time. The other five dealt with improving the process by reducing errors and eliminating redundancy. Implementing these recommendations reduced the development cycle from 39 to 24 months and improved both process effectiveness and efficiency. As the team became more familiar with the key elements of the process, the cycle was further reduced to 19 months a year later—overall, a 51 percent improvement. Cost savings in qualification testing and other process changes were estimated to be in excess of two million dollars. Considering the labor costs expended by the process team, the return on investment in the first year was 10 to 1, and it became considerably greater in subsequent years. The improvement in revenue of the end product by the cycle time reduction of the assemblies was not estimated.

The two most important improvements, considered breakthroughs or reengineering, involved (1) adoption of a modular design concept that allowed off-the-shelf use of standard functional circuit modules and (2) modification of the qualification strategy that enabled a Bayesian statistical approach to test time and sample sizes. In addition, a prequalification or exploratory test of the final design was introduced as a permanent improvement. By performing these discovery types of tests during the prototype phase, defects that would have caused reliability failures in the field or quality-type defects in manufacturing can be discovered early enough in development to readily modify the design, resulting in a more robust, higher quality product. Thus the concept of prevention was incorporated in process improvement.

The analysis also showed that this design process represented a microcosm of a much larger and complex development process having similar characteristics, constraints, and operational problems. Further, the use of this approach resulted in significant improvements in development time, process effectiveness, and efficiency and showed direct applicability to a larger, more complex development process. Several years later, the concept was applied to the development of the end product with results exceeding those obtained in this case.

Case 2: A Service Process: Methods for Improving Service Quality*

In a product development laboratory, services may range widely from cafeteria and personnel services all the way to support services, such as design release, and technical services. In this case, we examine another aspect of the development process: qualification. Qualification is a service-support type activity as opposed to design, which is considered an in-line activity. Qualification of a design varies widely both in its formalism and its complexity. It may range from simple us-

---

*This case was adapted from a paper presented by the author at ASQC Quality Congress in Minneapolis, May 1987; it appears in *AQC Transactions* under the title "Service Quality in an Engineering Laboratory."

ability tests for certain consumer products such as toys to large, elaborate, and time-consuming trials and experiments for such products as pharmaceuticals. In services there is an inherent customer-supplier relationship involved. The case analyzes qualification from a service process viewpoint.

Design qualification encompasses analysis and testing of a product to meet quality, reliability, safety, and, where applicable, industry-government standards. In this case, the product design to be qualified is an electronic assembly used for providing power to a high-reliability, high-end computer system developed by IBM. The new design is projected to have cost, performance, size, and weight advantages compared with older, conventional designs. The key question to be answered is whether certain technical objectives can be satisfied. The qualification process, through the employment of comprehensive evaluation and testing, addresses this question.

The functional format flow diagram of Figure 13-2 shows the activities involved. The process begins with a formal request by the developer to perform a qualification test. Agreements are made on schedules, test methods, and sample sizes, which become part of a comprehensive plan for approving the design. Subsequent activities involve execution of the tests and various evaluations that are conducted concurrently. For the most part, electronic assembly tests are of an accelerated nature and designed to verify in a relatively short time the long-term reliability performance of the product in service.

The major steps for qualifying the assembly include (in addition to accelerated testing) activities such as review and approval of the manufacturer's process and quality plan, application testing, electromagnetic interference testing, safety review, and Underwriters Laboratory and Canadian Standards Association review and approval. These activities were integral to the total qualification process. Failure to complete any one of them resulted in withholding the release of the design. Corrective action to remedy deficiencies in the design, components, materials, and so on was required to allow the release to go forward. Correction was initiated by the design engineering department, then reviewed and approved by both the qualification department and an independent product assurance function.

From a process management point of view, the "owner" of the development process (design, qualification, and release) is the product design manager. This manager is fully accountable for various aspects of the product and has the authority to make or approve changes—the primary aspect of ownership.

Owners of subprocesses provide a technical support service within the context of a customer-producer-supplier relationship (described in Chapter 3). This type of relationship is illustrated in Figure 13-3,

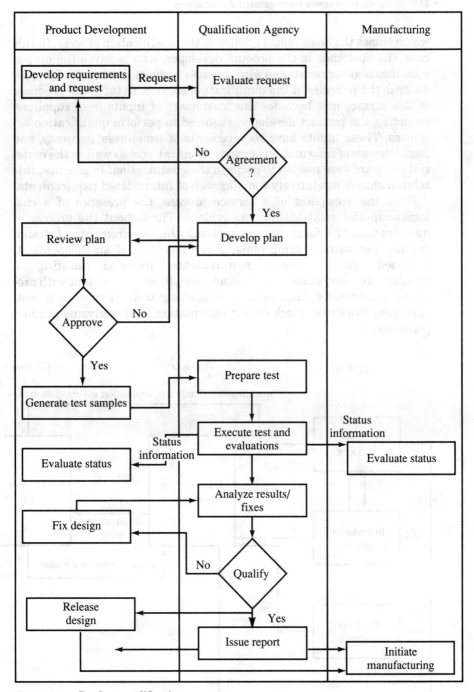

Product Development	Qualification Agency	Manufacturing

Develop requirements and request → Request → Evaluate request

Agreement? → No

Agreement? → Yes

Develop plan ← Review plan

Approve → No

Approve → Yes

Generate test samples → Prepare test

Execute test and evaluations

Status information → Evaluate status

Status information → Evaluate status

Analyze results/fixes

Fix design

Qualify → No

Qualify → Yes

Release design

Issue report

Initiate manufacturing

Figure 13-2.   Product qualification process.

which shows the basic model applied to the qualification process. In this case, the customer is the product developer, who receives the output (qualification approval) and also provides some of the inputs. The producer is the provider of the qualification service. In turn, the producer of this service now becomes the "customer" of inputs from suppliers, including the product developer, required to perform qualification activities. These inputs have requirements of timeliness, accuracy, and completeness of information (such as specifications) as well as the material and hardware needed to perform the qualification. In practice, this relationship is a relatively complex web of interrelated requirements.

From the viewpoint of a service process, the presence of a customer-supplier relationship was evident. Throughout the process of qualification, the focus was on meeting the requirements of design engineering without compromising the integrity of an independent, unbiased group. These requirements included meeting a schedule, devising optimum hardware sample sizes consistent with providing statistically valid results, conducting tests at minimum cost, supplying timely feedback of test information, and analyzing product performance.

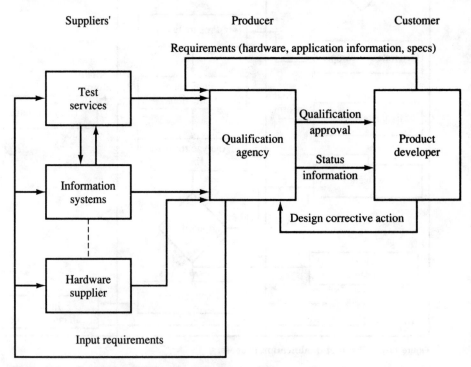

Figure 13-3.   Customer-qualifier-supplier relationship.

In service-support activities, a useful method for completely defining the various inputs and outputs between a customer and supplier(s) is a requirements matrix. The requirements matrix technique is discussed in Chapter 3. Figure 13-4 illustrates a portion of the matrix used for the qualification. The groups or functions participating in the qualification are indicated on one dimension of the matrix and the qualification activity required is listed on the other. A symbol or descriptive notation is placed at the intersection of the requirement

Activity	Design Engineering	Reliability Engineering	Materials Lab	Release Department			
				Responsible Organizations			
Qualification hardware	X 1						
Engineering specifications	X 2						
Qualification plan		X 3					
Reliability analysis		X 4					
Materials analysis			X 5				
Failure analysis	X 6	X 7	X 8				
•	•	•	•	•	•	•	•
•	•	•	•	•	•	•	•
•	•	•	•	•	•	•	•
•	•	•	•	•	•	•	•
Qualification approval		X M					
Release to manufacturing	X N			X N			

Commitments:

X1   Design Engineering provides quantity Y of qualification hardware by date Z_1.

X2   Design Engineering provides specifications by date Z_2.

X3   Reliability Engineering completes qualification plan by date Z_3.

Figure 13-4.   Requirements matrix for product qualification (partial).

to designate the group responsible for supplying it. For example, the design specifications needed in conducting a qualification are supplied by the design group; this is designated as X2 in the figure. Further information regarding commitments are given below the matrix.

The matrix approach is a useful technique for delineating the requirements and responsibilities involved in a service-support process; it also helps reduce the ambiguity existing at interfaces, an ambiguity that is prevalent in many business operations (as noted previously). In this case, it was found that many of the operational problems that occurred originated at the interface between one engineering department and another—traditional barriers that exist within a development organization. The requirements matrix that evolved created a greater understanding of what is required to execute the qualification and, hence, served to reduce or remove impediments to providing quality service.

There are two critical success factors of this service process: responsiveness and efficiency in detecting design errors. Four measures were instituted:

1. Response to a qualification request within a committed time interval.
2. Performance of the qualification within prescribed schedules.
3. Release approval within the required time period after satisfactory completion of qualification.
4. Deficiencies discovered during qualification testing.

Because of the discontinuous nature of qualification and because this was a job or project type of process, the parameters were combined into one measure called operational errors and reviewed monthly. The first three parameters represent a measure of the effectiveness of the service department. The fourth parameter is a measure of the effectiveness of the qualification process itself. These measures served as a basis for assessing improvement.

Reviewing the types of failures found during various qualification tests that were conducted on the assembly, it was found that many of them could have been prevented by earlier testing of prototypes. A prequalification test phase was then added to the development process to allow more timely evaluation of the design. This provided the capability of discovering deficiencies earlier in the process and fixing them prior to design completion. The end result was a significant reduction in test failures during formal qualification tests and a shorter testing cycle—another example of the payoff resulting from prevention improvements.

Other improvements to the design process included early manufacturing involvement to assure manufacturability of the design and earlier design reviews—both manifestations of concurrent engineering. The manufacturer also participated in the development process by fabricating the design prototypes. Performing preanalysis of the first-pass design proved beneficial in reducing the number of design changes that occurred. A comprehensive quality plan was put in place by manufacturing, giving the final product greater immunity from process variation and providing a more uniform, higher quality product.

Examining qualification as a service process provided an objective focus on the manner in which it was performed. By applying process management, substantial improvements to the development process were obtained, which served to reduce overall cycle time and total development costs as well as improve the robustness of the design. In addition, it was found that the customer-producer-supplier relationship model provided a sound model for improving service quality.

Case 3: Engineering Services: A Process-Team Approach to Improvement*

In this case, we examine how a team approach to analyzing the work product and its process resulted in significant and long-term improvement in the output quality of an engineering service department. The group involved is a printed-circuit design department in the IBM Product Development Laboratory in Kingston, New York. This department provides design automation support for printed-circuit cards and boards used in computer products developed in the laboratory.

Support involves the design and release to manufacturing of printed-circuit design data. These digital data, transmitted via telecommunication to IBM manufacturing locations, are known as TDI's (Technical Data Interfaces) and represent the digitized wiring pattern of a card. TDI's consist of two kinds of digital data: special (or custom) and standard, representing the two types of printed circuit cards designed in the laboratory.

The quality of the transmitted data is critical in this process, where circuit design information is used to create wiring patterns on a printed-circuit card or board. These patterns are converted to digital information of the two types mentioned previously. Prior to transmit-

---

*This case is based on "Teamwork—A Focus on Engineering Quality" by F. A. Mancuso in the 1988 *AQC Transactions* of the American Society for Quality Control, 42nd Annual Quality Congress, Milwaukee, 1988.

tal, the digital information is automatically compared against wiring criteria embedded in the computer checking programs. Designs containing wiring pattern discrepancies are rejected, then investigated and finally reworked into the proper digital data ready for transmittal. Errors of various kinds could and did occur in this process.

In 1979, the reject rate for standard designs in this operation was 27 percent; nearly one in every three designs required rework. The rate for custom designs was even higher, about 45 percent—a percentage not atypical for industry at the time. Over a period of five years, equipment and programming improvements reduced reject levels to about 5 percent. Nevertheless, continued improvement was felt to be not only desirable but necessary. Consistent with the company's focus on business processes, a strategy for further improvement was developed. The strategy, termed "revitalization" by the department manager, consisted of the department employees, organized as a team, examining the design process. The department manager, F. A. Mancuso, related the approach taken:

> The strategy used was one in which principles of process management were applied to analyze and improve the key processes of a business operation. Initially, frequent department meetings were held outlining new quality expectations and the employees' role in achieving zero rejects. A department process guide was planned where all of the department processes and procedures would reside. The process guide was to be a team effort and all department members would contribute.
>
> ... A department quality team was formed and met on a regular weekly basis. A team leader was elected by the group and was provided with additional education: Leadership skills, the role of quality teams in process improvement and knowledge of planning and running effective quality team meetings.
>
> The quality team first met in January of 1985. The entire department participated. Their first activity was to examine all department processes and procedures for deficiencies and problems that caused rejects. Their goal was to increase productivity, decrease cost and cycle time and provide high quality products and services to the customer. It should be noted that, while the business process was being revitalized, normal production was maintained.
>
> The team used the "brainstorming" method to identify problems and generate ideas for solutions. They then prioritized the list and resolved some of the procedural issues. These successes were all that were needed to spark the team for the real work ahead. A four-step quality plan was then developed: I) Define the process, II) Document the process, III) Establish measurements, and IV) Identify defects and their removal.
>
> All processes and procedures were reviewed by the team. Each one, as required, was rebuilt to adhere to the four-step plan. The team was now engaged in the revitalization of the department business structure. As

each process and procedure was completed, it was prepared for inclusion in the department process guide. During this period of high activity, there was noticeable enthusiasm in the behavior of team members.

Employees became self-motivated and caught-up in the revitalization movement. Daily production work was examined very closely. It was checked and rechecked by another department member to ensure completeness and accuracy. As problems arose and schedules were exposed, the team conducted emergency meetings to "brainstorm" a solution. They became reluctant to release a product unless it was 100 percent correct.

On completion of all the process reviews, a second pass was made focusing on measurements. Cost, cycle time and increased productivity were the focus items. All processes were examined for wasted steps and streamlined to be as effective and efficient as possible. Manual checks were minimized with the development of automatic program checking, further reducing errors. Because automatic program checking is inherent to the process, cycle time was decreased by 25 percent and productivity also improved. These improvements maximized computer usage to perform the iterative routing tasks which allowed more employee time to research ideas for additional improvements.

To further improve the department's effectiveness and efficiency and to examine the process boundaries, engineering input data was analyzed and found to be causing additional cost and cycle time. The team collected and analyzed this data and discovered we were making extra iterations through our process due to input errors and changes. Engineering was advised and education seminars were conducted to instruct them in all aspects of the department requirements. As a result, input improved and the department's engineering customers' cost and cycle time were reduced.

At the end of six months, in June of 1985, the department processes and procedures were completely rewritten and published in the department process guide. Over the next four months, rejects were further reduced. As rejects appeared, they were analyzed and resolved by the quality team to eliminate reoccurrence. The quality team goal of zero rejects was achieved in October 1985. The team worked a month at a time to avoid becoming overconfident and every new month without a reject increased their determination; they were resolved to maintain their record. After one year of zero rejects, they were honored at a luncheon hosted by senior laboratory management to mark the occasion. To date, after two years, the department output is still reject-free.

...The initial accomplishments were attaining zero rejects and building the department quality team. Long-term results include a revitalized department with an attitude of business ownership and a voice in how it is conducted. A quality awareness was also created, one of partnership with its customers, suppliers and manufacturing. The department understood engineering's product development environment, their own contribution and importance. The department also felt a sense of employee-management appreciation.

From January 1985 to December 1987, department members implemented $200,000.00 in cost savings ideas that reduced labor, materials and computer cost. Productivity and cycle time gains were also measured by method improvements which maximized computer resources. As the department became more efficient and effective, productivity increased. This allowed time for employees to participate in technical education, advanced technology experiments and new tool research.

The employee-management relationship improved significantly. From a management viewpoint, this was a satisfying aspect of this experience and an opportunity to influence change in both the employees and the business.[7]

This case demonstrated how a self-directed team taking a problem-solving, process management approach achieved a sustained defect-free operation with improved productivity.* The employees of the department, as a group, developed a sense of ownership of the process. Management clearly empowered the employees with process decision-making—normally a management prerogative. This was perceived and accepted by the members of the department and served to improve morale and build team cohesion. Again, the social component of the productive system was recognized as key in improving an operation.

In this chapter, we have described three process cases involving operations in an engineering laboratory. In Case 1, we saw how a process management analysis using a group approach examined the cycle time of producing a design. As an outcome of the analysis, ten key improvements were adopted and resulted in significant cost-savings. Five of these improvements enabled a reduction in the development cycle by 51 percent. The other five improved the overall effectiveness and productivity of the process. Case 2 shows how qualification, a subprocess of design, can be viewed as a service and examined both in terms of the customer-producer-supplier model and in terms of process management. A process examination resulted in the implementation of earlier testing of the design, which made for timely discovery of design deficiencies and application of corrective action. Finally, in Case 3, we saw how a process management approach using an empowered team led to defect-free output sustained over a substantial period of time.

In all three cases, it was important to adopt a total process view of the operation. To do anything different would have given only partial or piecemeal results. Process management has been shown to be an effective approach for improving the quality of the development pro-

---

*The manager, in a recent conversation, pointed out that the operation continued defect-free for well over three years. The case is currently used as an example in IBM of team empowerment.

cess. By applying it, one can examine how product development takes place. The very nature of the work-flow in product development can be addressed by viewing it as a process. Examining the characteristics of this work-flow provides a natural way to assess the effectiveness and efficiency of the work being performed. In turn, understanding the manner in which work is performed provides a sound basis for continuous improvement.

In Part II, we have described various case examples of applying process management in different business functions—administrative, financial, and engineering—in addition to a service operation. Several common elements emerged as we examined these divergent cases: absence of ownership, lack of definition, an absence of measurements and feedback, and little or no customer focus. Improvement was nonexistent. Processes were not being managed as processes; they became reactive. As a result, they lacked effectiveness and efficiency, customer requirements were not met, and waste existed.

These examples are not atypical of many of our business processes, as was pointed out in Chapter 1. When process management was applied, we saw significant reductions in error rates and cycle time and corresponding improvements in effectiveness and efficiency. A springboard for continual improvement of these processes was formed.

Notes

1. E. H. Melan, "Improving Responsiveness in Product Development," *Quality Progress*, June, 1987.
2. M. Kogure and Y. Akao, "Quality Function Deployment and CWQC in Japan," *Quality Progress*, October 1983.
3. L. P. Sullivan, "Quality Function Deployment," Quality Progress, June, 1986.
4. G. Box and N. Draper, *Evolutionary Operation: A Statistical Method for Process Improvement*, Wiley, 1969.
5. G. Taguchi, *System of Experimental Design*, American Supplier Institute Press and Unibup Kraus, 1987.
6. M. A. Verespej, "The R&D Challenge," *Industry Week*, May 4, 1987.
7. F. A. Mancuso, "Teamwork—A Focus on Engineering Quality," Transactions, 42nd Annual Quality Congress, May 1988, ASQC, Milwaukee, Wisconsin.

Processes:
Present and Future

Processes:
Present and Future

14

Designing a Process

Process management is generally applied to existing processes that have evolved over time. Often, administrative and service processes begin with a rudimentary set of activities contained within a work group or department. As the firm grows in size, these activities are merged into the various operations of the firm and grow into larger and more complex interfunctional processes. These complex processes are a product of evolution and adaptation from simpler sets of activities. Rarely are they designed from the beginning.

Business processes also evolve as a result of a response to some problem or a management-initiated change. The process is fixed, or "patched," by adding one or more activities to resolve the problem. The activity remains in place permanently long after the circumstances creating the problem have disappeared. An example of a business process that evolved from a management-initiated change is the Windsor Export case described in Chapter 12. Here, the accounts payable process that evolved was a response to a policy change that resulted in vendoring various internal activities that were performed in close proximity. The new process now comprised activities separated by thousands of miles, reducing its effectiveness.

Much of what we have described so far has been directed to processes that are already in place and functioning. However, the principles of process management can also be applied to new processes as well. In Chapter 2, various technical characteristics of a process such as transformation, feedback control, and repeatability were described that apply equally to established and to-be designed processes. These characteristics provided us with a better understanding of the nature of business processes. However, there are social and behavioral aspects that may affect its quality and productivity as we have seen in the cases described in the previous chapters. The landmark study by Mayo and Roethlisberger of an assembly operation at the Hawthorne works of Western Electric in Chicago in the 1920s was the first to point to the social aspects of a factory process and its influence on productivity. Prior to this, the prevailing approach to understanding and improving operations was the scientific management teachings of F. W. Taylor, which ignored the social element. Later, during the 1950s, action research performed by investigators such as Trist, Murray, and Bamforth at the Tavistock Institute in London, which became known as sociotechnical analysis, provided a greater understanding of the social aspects of work. In this chapter, we shall examine certain technical and social considerations in designing a process, the concept of work flow in process design, and a general model for constructing a process.

Requirements for Process Design

Like product design, process design is generally preceded and governed by certain requirements and objectives. Sociotechnical theory teaches that both technological and social components must be considered in designing a process. In order to achieve an effective, efficient, and adaptable process, these must be taken into account.

Traditionally, manufacturing, industrial, and systems engineering efforts are directed toward implementing technology in a process. The social component is often neglected or, at best, relegated to a secondary activity. Roethlisberger points out certain aspects of the social dimension that affect work:

> 3. . . . modern industry is built up of a number of small working groups. Between the individuals within these groups and between individuals of different groups, there exist patterns of behavior which are expressing differences in social relationship. Each job has its own social values and its rank in the social scale.
>
> 4. Each industrial concern has a social as well as a physical structure. Each employee not only has a physical place but he also has a social place in the factory. Any technical change on the part of management may therefore affect not only the physical but also the social location of an individual or group of employees. This fear of social dislocation is likely to be a constant threat to the social security of different individuals and groups of individuals within the industry.
>
> 5. The failure on the part of management to understand explicitly its social structure means that it often mistakes logical coordination for social integration. This confusion interferes with successful communication up and down the line as well as between different groups within the industry.[1]

Requirements and expectations may be classified as internal and external, or customer oriented. In terms of today's quality standards and customer orientation, a product or service process must first be governed by customer requirements. Once external or output requirements are understood, design parameters are then developed and the process is constructed. During its construction, internal customer requirements are taken into account.

Prior to the actual design, the following factors need to be addressed:

1. The objectives and purpose, or mission, of the process (or subprocess). At the outset, it is necessary to have a clear understanding of the primary output or deliverable, its attributes, and the recipient of this output.

2. In addition to understanding requirements of the product or service it is necessary to determine internal design requirements such as capacity, effective throughput rates, and timing as well as cost and quality characteristics. It is also desirable at this stage to enumerate the critical success factors governing this process.

3. The organization(s) encompassing the process. The process designer must have some understanding of the organizational structure and culture within the boundaries in which the process is to operate in order to determine the nature of the interfaces that will exist and the social features of the structure. Killman points out some of the social and cultural manifestations of boundary setting:

> Whenever a boundary is drawn around a small number of people, a group or team emerges which provides a strong psychological bond among its members; a "we" versus "they" attitude tends to reinforce the strength of the subunit boundary. The culture of each work unit also pressures each member to be loyal to the tribe and not to other work groups. Since members are usually in physical proximity to one another, the dictates and pressures of their work group are more powerful than any documents that "require" cooperation among all groups and loyalty to the whole organization. The latter may be just an amorphous mass, hardly competitive in loyalty with one's own peers and work group.
>
> In addition, most organizations design their performance appraisal systems so that subunit managers control the distribution and withholding of rewards. For example, subunit managers often have primary input (if not sole authority) in hiring and firing decisions, wage and salary increases, and promotion decisions. Such authority over important rewards to members tends to reinforce members' loyalty and adherence to their subunit's objectives more than to the organization's and certainly more than to some other subunit's objectives.[2]

As with organizational design, there are no simple rules for placing a process or a subprocess in an organization. In many cases, placement is governed by the technical system or technologies and skills employed, such as software development, actuarial work, data entry, or accounting. In other cases, activities of specialized work groups or departments are positioned by their mission and function within a firm: R&D, purchasing supplies and material, marketing, and so on. In the final analysis, any set of activities is positioned on the basis of mission requirements, functional or skill specialty, people availability, location, perceived importance, or the technical system itself. The concept of ordered work flow described in the next section provides another criterion for positioning.

4. The technologies to be employed in the process or that are available for use. For example, the use of computer equipment may dictate the need for data entry and verification activity in processes that require data translation from hard copy to an electronic medium. Another example is the use of bar coding equipment to provide inputs to a process. Benchmarking or competitive analysis may yield additional information on applicable technologies that may be employed in process design.

5. It is also appropriate at the onset to determine a process strategy to be employed. A useful concept that can be used in developing a strategy is the product-process matrix, which was developed for products but is applicable to some service operations.[3] In developing the matrix, Figure 14-1, we first note the range of output volume

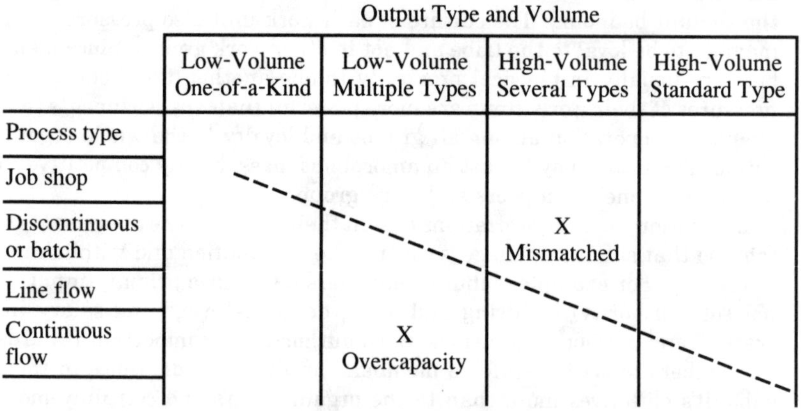

Output Type and Volume

	Low-Volume One-of-a-Kind	Low-Volume Multiple Types	High-Volume Several Types	High-Volume Standard Type
Process type				
Job shop				
Discontinuous or batch			X Mismatched	
Line flow				
Continuous flow		X Overcapacity		

Figure 14-1.   Product-process matrix. (Adapted from R. Hayes and S. Wheelwright, *Restoring Our Competitive Edge*, Wiley, 1984, p. 208.)

that the process must conceivably handle. Output may range from one of a kind or a few to very high volume of one or two types. The range is divided into three or four segments, depending on the mix of volume and output type, and is placed on the top row of the horizontal dimension.

Next, the type of process to provide the product or service is placed on the vertical dimension of the matrix in the left-most column. Processes can be divided into four major types: (a) job shop or intermittent processes in which one input (order) at a time is handled; (b) discontinuous or batch operation in which several inputs (job requests) are grouped together by type and processed as a batch in a repetitive fashion (here, output characteristics may vary widely); (c) line flow, where inputs of various types are processed as received through a set of dedicated transformation activities; and (d) continuous flow, where the process is high volume, output consists of a few standard types, and transformations are usually automated (chemical processes are prime examples of this).

In the product-process matrix of Figure 14-1, positions along the diagonal represent a perfect match of process to output type and volume. A particular process-output combination lying above the diagonal signifies that the process is mismatched with output requirements and, hence, may be inefficient and more costly. On the other hand, a process-output combination lying below the diagonal may indicate that the process may be capable of handling greater output than is actually demanded or anticipated. An over-capacity problem may exist, which can be costly—resources are not being applied to their full potential. In other words, the process is not operating efficiently.

Many administrative processes consist of activities involving both manual and automated tasks (as, for example, data entry and account verification). These consist of a combination of people and equipment. The limitation of many of these processes in terms of capacity and throughput is the human factor rather than technology. Capacity in these situations is increased by adding people. Increasing personnel at a service desk, for example, increases service capacity; the electronic information system containing customer data generally has capacity and response time far exceeding that of the total inquiry process. A designer must, therefore, take into account total process capability for some planned level of service output.

Another concept useful in developing a process strategy is that of reengineering. As noted in a previous chapter, reengineering is currently in vogue. Reengineering involves both innovative thinking and the application of information technology to business processes. Reengineering may be initiated by means of benchmarking—developing a knowledge of best-of-breed processes and applying it to a specific operation—or it may be self-initiated. An example of this is the accounts payable operation at Mazda. Ford management, in their search for ways of improving their accounts payable operation, discovered through their business relationship with Mazda that it was performing this function with only five people, in contrast to the several hundred employed by Ford. Upon investigating the reason for the marked difference in staffing, Ford found that Mazda processed payments without invoices. Payments were not initiated by a customary invoice, but rather by an acknowledgement from the receivers of the goods or services within the Mazda plant. This example underscores the fact that process design strategy must take into account the use of nontraditional, innovative methods and the questioning of old ways of doing business.

These design factors comprise the initialization phase of process design. From the beginning, the designer must have a clear idea of what the process is for and what it is intended to do, the organization(s) that contain it, the nature of ownership, and a strategy for developing the process. The designer must identify both the beginning and the end of the process as well as its requirements. An example of the preliminary design information needed is given in Table 14-1.

The Concept of Work Flow

Ideally, the most effective organization is one where no interfaces exist that detract from the quality of the work flow, where the output work product satisfies customer requirements, and where the suppliers satisfy the producers' requirements. In practice, this does not often happen.

In most business organizations, there is a hierarchical arrangement of layers of management reflecting authority and power—a command structure. The number of management layers, or the vertical depth, of any organization is a function of the size of the organization, criteria for span of control, organizational constraints, and the attitude of management. Some firms may have as many as ten layers of management between the employee and the CEO. Others of similar size may have only half that number. Deeply structured organizations tend to be highly bureaucratic, slow in making decisions, replete with communication problems, and inefficient (although they may appear from the outside to be well-managed).

TABLE 14-1   **Process design: boundary and
ownership information requirements**

1. General requirements
    a. Purpose and objectives of the process
    b. Primary output
    c. Secondary output (s)
    d. Design capacity
    e. Throughput rate (effective)
    f. Quality characteristics
    g. Critical success factors
2. Organizations and boundaries
    a. Primary organization(s)
    b. Support organization(s)
    c. The process ends with:
    d. The process begins with:
    e. Primary interfaces
3. Ownership
    a. Management
    b. Employee
    c. Team
4. Technologies to be used
5. Skill requirements
    a. Technical
    b. Interpersonal
    c. Decision-making

Regardless of the nature of the structure, however, work flows in a horizontal fashion from one worker to another across many departments and functions in an organization. In a product development organization, for example, a design may be created in one department while drawings and other paperwork come from another area and still another department will be responsible for the release of the design to manufacturing. The flow of design information proceeds in a somewhat sequential fashion from worker to worker across these departments until the entire design resides on hard copy or in some electronic data base.

On the other hand, information flow, communication of various kinds, and decision-making flows up, down, and horizontally within the structure. Because work flows horizontally, its quality depends on how each worker or work group performs the work itself. How well this is done depends on a complex array of factors, not the least of which are the organizational and interpersonal aspects of the work group. Roethlisberger describes some of these aspects:

> Industry has a social organization which cannot be treated independently from the technical problems of economic organization. An industrial or-

ganization is more than a plurality of individuals acting only with regard to their own economic interests. These individuals also have feelings and sentiments toward one another, and in their daily associations together, they tend to build up routine patterns of interaction. Most of the individuals who live among these patterns come to accept them as obvious and necessary truths and to react as they dictate.

If one looks at a factory situation, for example, one finds individuals and groups of people associated together at work, acting in certain accepted and prescribed ways toward one another. There is not complete homogeneity of behavior between individuals or between one group of individuals and another, but rather there are differences of behavior expressing differences in social relationship. Individuals conscious of their membership in certain groups are reacting in certain accepted ways to other individuals representing another group. Behavior varies according to these stereotyped conceptions of relationships. The worker, for example, behaves toward his foreman in one way, toward his first line supervisor in another way, and toward his fellow worker in still another.... These subtle nuances of relationship are so much a part of our everyday life that they are commonplace.[4]

The notion of work flow is of primary importance in process design. The interface created at the intersection between the horizontal flow of work and the vertical structure of organization tends to detract from the quality of a process.

In the early 1960s Chapple and Sayles examined the nature of work flow in firms and proposed that industrial organizations be based on ordered work flow. Organizations, the authors claimed, should be based on a logical flow of work rather than on the traditional functional groupings based on specialization:

Most decisions on "proper" organizational structure are based primarily on similarities of activities or functions. Traditionally, organizations are divided into such major functions as sales, production, finance and personnel. Each may have subsidiary functions such as engineering, training, market research, inventory control, etc. that, in many companies, compete for equal standing with the others...

The endless arguments about the proper placement of organizational activities are usually only temporarily settled. Because only verbal criteria exist, no one can define a function accurately and no one wins a conclusive victory...

Clearly, a different approach to the problem of organizational design is needed. The structure built for members of management can be ignored for the moment to go back to the bottom where the work is done.

This requires looking at the way technology separates a series of jobs that must be accomplished if the product is to result. We may manufacture something, buy it for resale, or hire it, as in the case of money, but whatever the business—manufacturing, retailing, banking or service— we follow certain techniques. There is a beginning, when the process starts, something is done, and the process ends.[5]

Every organization, they point out, has a method of performing work that involves some sequence of operations. In delineating work flow, the first step is to identify units of work. The next step is to set meaningful work boundaries under a single supervisor encompassing a logical set of work units. In doing this, we reduce the potential conflict between the classic, vertical organizational structure and the flow of work. By setting process boundaries under a single manager, we eliminate various suboptimization effects that are encountered because of the conflict between functions having diverse goals. Doing this gives the organization a "flatter" structure and promotes more effective and efficient work flow. Because of competitive forces, many firms today are examining this approach to restructuring organizations. Flatter organizations, on the other hand, tend to force management to address the empowerment issue. The increased span of control that occurs reduces contact time between the manager and the employee and places greater emphasis on employee initiative.

Constructing a Process

A process designer, then, must take into account the organizational, social, and work flow aspects of an operation. In some cases, a process is almost totally contained within a few departments in a function; in others, it must, of necessity, cross functions. Where work flow proceeds across functional entities, a practical consideration for the designer to keep in mind is that process effectiveness can vary inversely with the number of interfaces that the work is required to flow through. The greater the number of interfaces, the less likely will the process be effective because each interface creates a potential for misunderstanding and miscommunicating work flow requirements.

Where the designer has to deal with a set of interfaces that exist by virtue of the organizations encompassing the process, developing an understanding of the work flow requirements of each interface becomes important. Applying the customer-producer-supplier relationship model (discussed in Chapter 3) at each interface enables the designer to define output characteristics in terms of customer requirements. In turn, the activity steps needed to provide this output are developed, which, in turn, leads to defining the input requirements for the work that the preceding activity set must provide. In this manner, the designer proceeds in sequence from the primary deliverable at the output boundary, working back through the various outputs at each interface until the input boundary is reached. Each activity represents one or more transformations needed to convert inputs to some required output. The designer may use one of the methods described in Chapter 3 to define the interface requirements.

Hierarchy model

A useful construct to employ in designing a process is a hierarchy model. Based on the initial information gathered by the designer (Table 14-1), a macro flow diagram of the subprocesses composing the process can be constructed (Figure 14-2). In other words, the designer draws a high-level model of the process to delineate the output and input boundaries, the key interfaces, and the major sets of activities. From this high-level flow, the activities of each subprocess can then be designed using the CPS model and the technology information obtained during the establishment of process requirements (part 4 of Table 14-1).

Prior to developing the high-level flow model, the process owner must be identified. The designer must not only keep the owner continually informed of the progress of the design but also take the owner's requirements into account in developing the design. The designer must also understand any differences that may exist between customer's and owner's requirements and be able to resolve them in a timely manner.

Categories of activity

The next phase of the design involves detailing the activities needed to satisfy each subprocess output requirement. Activities can be divided into three categories in terms of work flow: sequential, parallel, and reciprocal. Sequential activities occur when one activity must be completed prior to the start of another. Parallel, as the name implies, means that two or more activities can be conducted independently and at the same time. Reciprocal flow occurs when a two-way interaction occurs among individuals involved in two or more separate activities. Figure 14-3 illustrates the three types of activity flow. It is important for the designer to keep these three types in mind during the design, because each type varies in the degree of coordination needed in managing the work flow.

Thompson[6] points out that parallel flows require less coordination than the other two because the outputs are independent and can be readily combined by instituting rules and procedures. For example, in

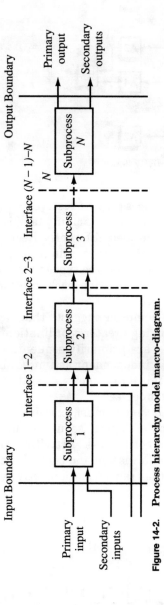

Figure 14-2. Process hierarchy model macro-diagram.

239

Sequential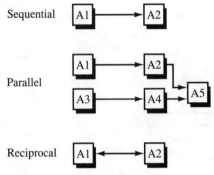

Parallel

Reciprocal

Figure 14-3.   Types of activity flow.

accounts payable, invoices may be received either by mail or FAX machine. In either case, the time-stamping task is the same. However, the log-in task may be performed differently. Reciprocal flows, he suggests, require the greatest degree of coordination and, hence, are the least efficient and costliest of the three types to manage. This type requires frequent recycling of information back and forth until a satisfactory work output is achieved.

For this reason, Thompson suggests that it is more efficient to include activities requiring greater coordination within a single subprocess. These activities are primarily sequential and reciprocal. Where possible, parallel and independent activities should be arranged so as to cross interfaces. Figure 14-4 shows several of these types of activities arranged within subprocesses.

At this point, the designer must examine the set of activities being constructed in terms of logical sequence, potential error or failure points, and capacity bottlenecks. Activities to be designed in are of two basic types: transformation and control. Control governs the quality and stability of the transformation activity; transformation is the

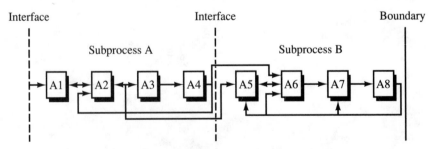

Figure 14-4.   Activities within and between subprocesses.

added-value characteristic of the process. Control points serve to reduce or eliminate errors or failures resulting from the transformation and may also serve as measurement points. Since they are non-value-add, the designer must use and position them judiciously, keeping in mind that there is a cost to inspection as well as a cost in propagating an error. The volume of work flow must also be understood in order to assess the resources required to execute transformation and control. For this, activity execution times as well as performance and allowance factors are needed to establish standard times and, in turn, the capacity for each activity.

Many of the process definition steps described in Chapter 4 are applicable to laying out activities. These steps include defining the

- Output(s) for each transformation
- Work activity needed to perform the transformation and provide the output(s)
- Inputs necessary to perform the transformation

In this manner we can proceed to lay out the process from the final output to the first set of inputs needed to initiate the first activity. For many service and administrative operations it is beneficial to proceed from output to input. However, in the case of product processes involving many physical transformations based on technology it may be more appropriate to design the process from the input side to the output. Here we determine the first activity performed as the inputs cross the input boundary and proceed in sequence from one unit operation to another until the final activity producing the output is reached.

Having completed the activity-level design, the designer must now focus on the tasks composing each activity. Each activity can be subdivided into two or more tasks. Using criteria such as value-add, timing, and control, the designer must devise the kinds of tasks to be performed for each activity. Tasks are also divided into two types: transformation and control. In transformation, we convert inputs to something of a higher added value. Control, on the other hand, involves inspecting or verifying, reconciling, and decision-making (as a consequence of control). Decision-making may result in a corrective transformation (such as rework, repair, and modification). These can be deemed non-value-add work that is the result of an error.

As an example, in accounts payable, the major activities on the input side are receipt and validation of the invoice. Tasks involved in the receipt of the invoice are time-stamping (to record the date and time the document was received) and logging-in the document. Time-stamping may be performed for both measurement and control purposes. For timing, the organization may want to take advantage of discount terms of payment. For control purposes, then, timeliness of payment may be a critical success factor of the accounts payable process. Noting the date of receipt of an invoice as well as the issuance of the payment are tasks that provide a means to control payment response time.

Upon completion of the task-level design of each subprocess, the designer must examine each subprocess in terms of design capacity, effective throughput, quality characteristics, and critical success factors (part 1 of Table 14-1) to determine the type and frequency of measurements that must be performed, the degree of data automation, and the staffing required. From this, process costs can be estimated.

Process costs

Process costs may be gotten by adding the total costs of all of the activities contained within the boundaries. These would include labor, material, equipment, and support costs (such as information systems). Once having the total cost, we can compute the cost of activities that are purely value-add, namely, transformation activity, and determine a value-add ratio, V_a:

$$V_a = \frac{C_{VA}}{C_T}$$

where C_{VA} is the total value-add activity cost and C_T the total process cost. A low V_a ratio is generally indicative of many non-value-add activities required to support the transformation and provide control. A high ratio, on the other hand, may be indicative of an efficient process having few non- value-add activities, or it may indicate an inherently high cost of transformation (as in the case of capital-intensive operations). The value-add ratio can guide the designer to examine the process in terms of activities that create added value and those that do not. In turn, non-value-add activities should be assessed against process requirements to determine if they are necessary.

Human factors

The technical and interpersonal skill requirements of the various activities comprising the process must be analyzed and understood. To address the social dimension of the process, it is important to know whether activities require team or individual work skills.

For some designs, a simple table enumerating the activities and tasks to be performed and their technical and social demands is sufficient to determine the social requirements of the process. In other cases, a dependency matrix may be appropriate and can serve as a means for determining these requirements.

Herbst[7] has developed a task-dependency matrix based on four key elements of a work relationship structure. These are:

1. *An activity relationship.* An individual can conduct an activity together with one or more individuals or separately. For purposes of developing a dependency matrix, the former is designated as a and the latter as $\bar{a}$.
2. *A task dependency.* Tasks that an individual conducts that are dependent on another are designated as d^*. Interdependent tasks between individuals 1 and 2 are shown as d_{12}. Independent tasks are designated as $\bar{d}$.
3. *Differentiation of worker role.* Tasks that are identical with another worker's are designated as i. Ones that are different are shown as $\bar{i}$. Tasks that are different but are rotated among workers are shown as i^*.
4. *Goal dependence among the individuals.* Goals can be either shared (g) or independent $(\bar{g})$. Goals that are nonreciprocal (i.e., noninterdependent) but supporting between two individuals, 1 and 2, are designated as g_{12}.

Individual matrices are developed based on a task analysis of a group, one for each of the four elements. The four matrices are then combined into one task matrix. Consider, for example, three people designated 1, 2, and 3 working on a set of activities in a process. 1 and 3 work together but have different goals; 3 is task dependent on 1. Individual 2 works separately and has different tasks from both 1 and 3; 3 also has different goals from 1 but has shared goals with 2. Individual 2 is task interdependent on 3 but not 1. The resultant combined task matrix is shown in Figure 14-5.

This matrix provides a relationship-dependency map among the individuals working in a process and provides a method for noting these relationships using symbols. Activities requiring a high degree of interdependence, task relationships, and goal dependence are prime candidates for a team approach to process implementation. In contrast, for processes having many independent activities and task relationships,

		Worker	
		3	2
Worker	1	$ad_{1,3}\bar{i}\bar{g}$	$\bar{a}d_{1,2}\bar{i}\bar{g}$
	2	$\bar{a}d_{3,2}\bar{i}g_{3,2}$	—

Figure 14-5. Combined task matrix.

individual workers empowered with some decision-making responsibility may be appropriate.

It is also important for the designer to understand the degree of empowerment of the group or individual from the viewpoint of executing responsibility (ownership) and decision making. Empowerment is based on the management culture and philosophy of the organization, its sociotechnical objectives, and individual management style.

Design completion

As the design is being finalized, it is appropriate to review all aspects of the design with the owner, including any open issues. After satisfactory completion of the review, attention must be given to implementation. Equipment and personnel previously selected to meet technical and social requirements are now assigned to the process. Training, either on the job or in the classroom (or a combination of both) is conducted. Task and activity level flowcharts are excellent sources of training material. Other descriptive documentation (such as that described in Chapter 3) is also useful for training. A technique to create employee involvement is having the staff assigned to the sub-process write the task and activity descriptions after a suitable period of time to gain familiarity with the work. The designer should reevaluate the social dimensions of the design to verify that interpersonal and group factors have been taken properly into account. Face to face interviews, either individual or group, and written opinion surveys for participants, customers, and suppliers can serve as means for eliciting feedback on the adequacy of the design.

In-process and output measurements serve as statistical information for the designer as well as the owner. This information is compared with the original requirements in terms of throughput, response time, and other design specifications. It also serves as a means for determining the effectiveness and efficiency of the process as well as providing an assessment as to whether or not the process is capable and in control. Evaluation criteria, described in detail in Chapter 8, are useful for assessing the design.

As we have noted, designing effective and efficient processes requires consideration of both its technical and social dimensions. In this respect, the traditional approach of making people fit the process is not only outmoded but leaves much to be desired. The quality of a process is not merely a function of how well its technical aspects are executed. A congruence of process, organization, people, and environment must exist if a process is to be effective and efficient. Both the designer and owner must take this fact into account in the design and management of processes.

Notes

1. C. R. Walker, *Modern Technology and Civilization*, McGraw-Hill, 1962. Reprinted with permission, McGraw-Hill, Inc., ©1962.
2. R. H. Killman, *Beyond the Quick Fix*, Jossey-Bass, 1984.
3. R. H. Hayes and S. C. Wheelwright, *Restoring our Competitive Edge: Competing Through Manufacturing*, Wiley, 1984.
4. Walker, *op. cit.* pp. 186–187.
5. E. D. Chapple and L. R. Sayles, *The Measure of Management*, MacMillan, 1961.
6. S. D. Thompson, *Organizations in Action*, McGraw-Hill, 1967.
7. P. G. Herbst, "Task Structure and Work Relations," Tavistock Institute of Human Relations, London, TIHR Doc. No. 528 (1959).

15

Future Trends

The Quality Revolution

In the 1970s the United States was rudely awakened to the superior quality and increasing market share of imports and the decline of domestic manufacturing. The 1980s witnessed a quality revolution in the face of economic decline in an increasingly competitive world. We have seen growing interest in the teachings of the quality gurus such as Crosby, Deming, Feigenbaum, and Juran—all Americans. Concepts such as quality improvement, statistical quality control, and total quality management, although not new, became widely publicized in the past decade. Improvement as a formal, managed program has become widespread.

An increasing number of organizations in the public and private sectors have embarked on quality improvement programs—some in the hope of improving their competitive posture. In 1988, a survey of 130 firms showed that nearly 50 percent have quality improvement programs.[1] These companies vary in their approach to improvement, using a mix of internally and externally developed methods. All industries surveyed predicted continued growth of these programs. A more recent survey by the American Management Association of 528 corporate executives showed that 91% of the U.S. companies they represent now have formal quality improvement programs. Of these 69% have given quality improvement high priority.[2]

In the beginning of the 1990s, we have seen a growing momentum of the quality revolution and increasing interest by government and education. The federal government, through the Department of Commerce, is promoting industry-wide competition in quality improvement by sponsoring the Malcolm Baldrige Award. An increasing number of firms are showing interest in the competition by applying for the award and many are using the award criteria as a means for assessing their quality systems. The government has established the Federal Quality Institute to provide the basics of TQM, start-up services and implementation assistance to various agencies of the executive branch.[3] State and local governments are also beginning to show interest and involvement in quality improvement. New York State now sponsors an Excelsior Award for quality patterned after the Malcolm Baldrige Award. The city government of Madison, Wisconsin with the assistance of the University of Wisconsin has begun quality improvement in its services. Numerous quality networks encompassing business, education, and government within a geographic region have been established to share and disseminate information on quality improvement and provide initiatives in various aspects of implementing TQM.

Education has finally awakened as well. There are signs of growing involvement of both K-12 and higher education. Schools from Alaska to Florida are beginning to examine ways of incorporating elements of TQM into teaching and into school operations. Mount Edgecumbe High School, Alaska's only public boarding school is applying some of Deming's teachings to student education and provides training in TQM techniques. An increasing number of colleges and universities are beginning to examine the concepts and methods of TQM, develop TQM courses and course modules, and apply them to the operations of their institutions. Fordham University offers an MBA in Total Quality Management. The University of Pennsylvania is using the hoshin planning technique to develop a 5-year plan for the institution and has adopted the process management approach to improvement.

The quality revolution shows no signs of abating.

As noted in Chapter 10, a single, unique approach to improvement does not and may never exist. Experience has shown that successful and sustained improvement requires four key elements:

- First, ongoing commitment and involvement by management— in a word, leadership.

- Second, an approach and strategy consistent with organization culture and values.

- Third, continuous motivation and participation by employees.

- Fourth, a system to sustain improvement on an ongoing basis— continuous improvement.

These elements are recognized to be of such fundamental importance that they have been embodied in the criteria for the Malcolm Baldrige Award. The quality system implied in the Baldrige criteria involves an active management role, an active employee role, a quality improvement strategy for the firm, a method of implementation, and a means for measuring results—all basic elements of any quality improvement system.

Without these elements and without a means of providing continuity, improvement will be sporadic at best. Continuous improvement has to be translated into tactical strategy for effective results. Process management provides a unifying methodology for continuous improvement in every function of the firm because processes are the common denominator of a productive system.

Organizational Trends

Yet, in spite of the quality revolution that is occurring, America continues to lose its competitiveness in many business areas. Although there are various social, political, economic, and technological reasons for this, one of the prime reasons remains bureaucracy as we noted in the beginning of this book. Richard Rosecrance, a professor at the University of California, observes:

> We must recognize that most of our organizations—public and private, military and economic—have too many chiefs and not enough Indians. We have embraced Max Weber's bureaucratic society with a vengeance with overstaffed corporate headquarters in which specialization has been narrowed to a pinpoint. Committees proliferate, but the job does not get done. It is almost as if leading American businesses were run by a series of ganglia but not by a single brain.
>
> More than half of the modern American corporations consist of workers uninvolved in operations or production work, an astounding fact. At the General Motors Corporation, 77.5 percent of the work force is white-collar and salaried while only 22.5 percent are hourly blue-collar workers. At Mobil Oil, 61.5 percent of the staff is white-collar; at General Electric, 60 percent, at Dupont, 57.1 percent...The ratio in typical corporations in Japan is about one-sixth of the American figure.[4]

Although the figures seem high, it must be remembered that firms with a high degree of automation, complex information systems, and substantial R&D efforts will have higher percentages of indirect or support personnel than direct or blue-collar personnel. Most of these are the knowledge workers of today. Nevertheless, an important way to improve competitiveness is to address the bureaucratic structure of the firm and its byproducts of suboptimization, high costs, and poor productivity. Rosecrance states:

> The industry of the future needs fewer but much more broadly trained executives. Workers should take on some management tasks. Many companies are already experimenting with self-managing work teams.
>
> If the United States is to arrest its decline, two developments need to occur in the near future: First, American business will need to increase the ratio of operators to managerial staff, giving management a much wider purview; second, workers will increasingly come to fulfill a management function.

As noted in the previous chapter, there is increasing recognition that "flat" organizations with few management layers are more efficient and productive than the traditional, vertical structures. James R. Houghton, the Chairman of Corning Glass, writing in the *New York Times,* exemplifies management thinking about new organizational roles:

> We have found that the successful operation of a global management network requires a new mind set. A network is egalitarian. There is no parent company. A corporate staff is no more or less important than a line organization group.
>
> A critical question we are examining is the role of people at the center of the network—the traditional corporate staff.... The hub will assume some of the functions of a broker, conflict negotiator, facilitator and think tank and that its major role will be to provide the information, resources and guidance to keep the network functioning.
>
> Today's better-educated workers want to have the power to control their own workplace lives. This should be encouraged. In our factories and businesses, we have hundreds of teams that spot trouble and fix it at the source without supervisors and top managers, like me, interfering.[5]

It has taken a century for management to come full circle in its thinking about the structure of a firm and how work is to be performed. We now recognize that addressing structure alone will not result in an efficient, competitive organization. A business is a complex ensemble of interdependent activities organized into groups. People who work in groups tend to lose sight of the whole and concentrate on the specific activity that they are paid to perform. When this occurs, the result is predictable—even in small organizations:

> Virtually everyone had the same perception of the organization and its problems. They described the organization as being composed of small, isolated groups, which they referred to as "clumps." The term *clump* was so prominent in people's minds that several remarked with frustration that their organization practiced the "clumponian" theory of man-

agement. People in the various clumps communicated with one another with extreme difficulty and, at best, were loyal to their clump. They had little sense of belonging to their company let alone the parent corporation. Individuals with 20 years tenure admitted to not knowing the names of coworkers in this small organization, even those who inhabited the same halls.

In reality, the company was an organization in name only. It was fractionated along very tight, specialized lines. People were suspicious of, and hostile to, one another and there was a dearth of communication between the various functional groups whose coordinated energies were required for effective function in the marketplace. As a consequence, this division was unable to devise, produce or sell the products that increasingly sophisticated customers demanded.[6]

Eliminating "clumponianism" is clearly the responsiblity of management. Without the adoption of the systems approach to the enterprise and the use of organizational development and team-building concepts, a firm will continue to be mired in clumps and suffer the consequences of poor productivity and low employee morale. Deming's dictum of breaking down the barriers is most appropriate for fragmented, compartmentalized organizations. Management must not only develop the organization in terms of its external relationships but also focus on its internal operations and the way people work together. Managers who concentrate on the external aspects of the organization and neglect its internal functions have only themselves to blame for productivity and morale problems.

According to Mitroff,[7] two distinct strategic trends are developing to improve a firm's competitive posture: a focus on the internal functioning of the company and a search for new economic relationships.

Internal focus

Internal focus encompasses process, workers, managers, outsourcing, antibureaucratic concepts, and productivity tools. Improving the processes of an organization has been the subject of this book. Process improvement, however, is not yet prevalent in this country, as the International Quality Study has shown.[8] This study discovered that only 12% of businesses in the United States use process simplification consistently as a means to improve business processes. In contrast, nearly half of the Japanese businesses in the survey use it more than 90% of the time for improvement.

In spite of the lack of attention of American business to process improvements, the importance of making such improvements is underscored by the emphasis on process in the Baldrige Award criteria. The 1992 guidelines show "Management of Process Quality" as one of the seven examination categories to be addressed in applying for the award. The preface to the category description states "The *Management of Process Quality* category examines the systematic process the company uses to pursue ever-higher quality and company performance. Examined are the key elements of process management, including design, management of process quality for all work units and suppliers, systematic quality improvements, and quality assessment."[*] This category is divided into five sections:

1. Design and introduction of quality products and services. Included are areas such as how products, services, and processes are developed, how designs are reviewed and validated, and how designs are evaluated and made more effective.

2. Process management of product and service production and delivery processes. Areas to address are how the company maintains process quality, how processes are analyzed and improved, how performance data are analyzed and translated into improvement, and how the company integrates process improvement with day-to-day process management.

3. Process management of business processes and support services. Areas to be addressed in the application are how the company maintains the quality of its business processes and support services, and how it improves them to achieve better quality, performance, and cycle time.

4. Supplier quality. Included are the methods the firm uses to define its quality requirements to its suppliers, methods used to assure that requirement are met, and the actions taken to improve the quality and responsiveness of suppliers.

5. Quality assessment. Involved here are the approaches the firm uses to assess (a) systems, processes, and practices and (b) products and services. Included is how the assessment findings are used to improve (a) and (b).

Worker-focused efforts include employee involvement, quality improvement teams, and task redesign by means of self-directed work teams (as, for instance, the Saturn project). An example of a successful employee involvement program is that of the A. O. Smith factory in Milwaukee, an auto frame manufacturer:[9]

---

[*]*1992 Award Criteria,* Malcolm Baldrige National Quality Award, p. 22, National Institute of Standards and Technology, Gaithersburg, Md.

Smith's Automotive Products Company embarked on an odyssey of work-place changes—hoping to revitalize itself through employee involvement (EI). The company and its unions found that low-level EI programs such as quality circles—off-line discussion groups that suggest ways of improving quality and cutting costs—couldn't reform Smith's hoary work practices. That would require radical changes in how work is organized and managed.

It took six long years. But finally, in 1987, Smith took the big step and began reorganizing workers into production teams that, for all practical purposes, manage themselves. The results were dramatic: In 1988, the productivity growth rate doubled and defects on the Ranger line were down to 3%.

The radical change that the firm undertook was to establish empowered work teams.

The work teams consist of five to seven workers who rotate from job to job. The members elect team leaders who assume many managerial duties such as scheduling production and overtime, ordering maintenance work and stopping the line to correct defects. Most rank-and-file workers don't object to the added duties and their increased power on the shop floor is no illusion. They even get to revise work standards set by engineers. "They just turned control of the shops over to us," says Steward Charles Perkins. "Our destiny is in our own hands. I love it!"

These work teams essentially replaced first-line management. The effect of these self-directed work teams on supervisory management was severe:

With hourly workers taking over managerial tasks, Smith retired or fired scores of first-line foremen, reducing the ratio of foremen to workers from 1 to 10 in 1987 to the current 1 to 34. While this reduction enabled Smith to cut overhead costs, it also undermined the morale of the remaining supervisors. Belatedly, the company is now training them to put aside their old management-by-control methods and adopt a participative style that suits Smith's new culture.

Another aspect of internal focus has been the sheer downsizing of organizations by eliminating layers of management and substantially increasing the span of control of the remaining managers. Staffs are also being drastically reduced. The business literature is replete with articles on eliminating middle management and staff in various Fortune 500 firms such as GM and IBM. IBM has eliminated two layers of management from its hierarchy and has decentralized its organization structure to reduce its decision making inertia. Even the U.S. Postal service, a reputed bureaucracy of 748,000 people plans to eliminate 30,000 managerial and supervisory personnel including nearly 50% of its headquarters executives through early retirement.[10]

Staffing is also being reduced by contracting out support services such as cafeteria, maintenance, library, purchasing, and data entry. Outsourcing, although not a new strategy, is increasing. It represents a lower labor cost method of doing business and creates greater opportunity for vendor services. Although its main use has been for manufacturing items, administrative work and services are also being outsourced. For example, a financial firm in New York vendors its back-room accounting and data-entry operations to a firm in Ireland.

Companies, particularly those encumbered by bureaucratic rules, have found that, by establishing separate business units or "tiger teams" unconstrained by bureaucratic ground rules, new products can be developed and manufactured in a fraction of the time. A technique called concurrent or simultaneous engineering also serves to reduce the design cycle. Functions such as manufacturing, procurement, field service, and marketing jointly participate with design engineering in developing the product virtually from its inception, thereby reducing sequential activities and improving the product's manufacturability.

Labor-saving tools to improve productivity have also become a significant part of the internal focus. Computer-aided design tools such as CADAM and CATIA, automated manufacturing tools such as robotics, methods such as just-in-time and focused factories, and information systems such as MRP are examples. Manufacturing management has taken greater interest in the concepts embodying "lean production" as a means for improving competitiveness through lower operating costs.[11]

Increasingly, more and more companies are adopting these approaches. Large firms such as General Electric, General Motors, AT&T, IBM, Johnson and Johnson, DEC, and Proctor and Gamble, as well as numerous medium- and smaller-sized firms and nonprofit institutions are now employing many of these methods as a means for reducing operational costs and improving efficiency.

External focus

The second trend, a search for new economic relationships in dealing with the business environment, has led firms to develop new associations. Joint ventures and alliances have become the strategy for sharing scarce and expensive resources (as in the case of technology skills) as well as risks. Joint ventures of U.S. businesses with both Asian and European firms are common, particularly where high capital investment is required. Capital-intensive operations such as semi-conductor technology is a case in point. Business arrangements previously thought improbable have now been formed: IBM and Toshiba, IBM and Apple Computer, Chrysler and Mitsubishi, Ford and Mazda and many others now exist. Consortiums such as Sematech have also been formed for many of the same reasons as well as to preserve technology within a nation.

This search for new ways of surviving knows no national bounds. Industry has become increasingly global and so has competition. Accompanying the one-on-one type of business alliances and joint ventures with foreign firms are the multinational trade agreements such as GATT and economic communities such as the ECC that serve both to promote and limit trade.

A New View

It is evident that the 1990s will see business operations being conducted by new paradigms. There is increasing realization that traditional approaches to management organization and competition are obsolete in the new global environment of business. The old authoritarian command structure of management as viewed by Weber is no longer viable in an increasingly technological world where information has become a primary work product. The classic responsibilities of management espoused by Fayol are being questioned. An era of enlightenment has begun.

Many traditional, command-type organizations will eventually be replaced by flat, decentralized power structures, classic management responsibilities will give way to empowered workers, and managers will become coaches rather than leaders in the traditional sense of the word. The new breed of manager and employee will need to take a systems view of the organization to manage and work effectively. Quality will become an integral part of daily work. Quality will transcend the traditional definition of merely meeting customer expectations and will come to mean exciting the customer—providing products and services that exceed expectations.

An insight into this new world is provided by Peter Drucker's view of the new manufacturing organization of the 1990s:

> In the traditional plant, each sector and department reports separately upstairs. And it reports what upstairs has asked for. In the factory of 1999, sectors and departments will have to think through what information they need from whom. A good deal of this information will flow sideways and across department lines, not upstairs. The factory of 1999 will be an information network.
>
> Traditional approaches all see the factory as a collection of individual machines and individual operations. The nineteenth century factory was

an assemblage of machines. Taylor's scientific management broke up each job into individual operations and then put those operations together into new and different jobs. "Modern" twentieth century concepts—the assembly line and cost accounting—define performance as the sum of lowest cost operations. But none of the new concepts is much concerned with performance of the parts. Indeed, the parts as such can only under-perform.

Consequently, all the managers will have to know and understand the entire process, just as the destroyer commander has to know and understand the tactical plan of the entire flotilla. In the factory of 1999, managers will have to think and act as team members, mindful of the performance of the whole.

Even managers with no business responsibility . . . will have to manage with an awareness of business considerations well beyond the plant.[12]

What Drucker espouses, then, is that managers must adopt a systems-process view of their organization—an integrated view of their world. Managers and employees alike must begin to view their operation as a component of a productive system.

In all of these developments there is one constant: products are produced and services are performed by means of processes. Business is a complex productive system composed of interrelated processes. For a business to be competitive, its processes must be effective and efficient. Process management provides a means to remain competitive in today's world. In Drucker's words,"the process produces results."

Notes

1. Goal/QPC Research Report No. 90-04-01, Goal/QPC, Methuen, MA.
2. Survey Performed by Ogilvey Adams & Rinehart for the AMA 1991; American Management Association, New York.
3. B. Stratton, "Federal Quality Missionaries," *Quality Progress,* May 1991.
4. *New York Times,* July 15, 1990. Quotation reprinted by permission, The New York Times Company, Copyright 1990.
5. *New York Times,* September 24, 1989. Quotation reprinted by permission, The New York Times Company, Copyright 1989.
6. S. A. Mohrman and I. I. Mitroff, "Business *Not* As Usual," *Training and Development Journal,* June 1987. Quotation reprinted with permission, the American Society for Training and Development, Copyright June 1987, all rights reserved.
7. I. I. Mitroff, *Business NOT As Usual,* Jossey-Bass, 1987.
8. *International Quality Study,* A Joint Project of Ernst & Young and the American Quality Foundation, New York, 1991.
9. "The Cultural Revolution at A.O. Smith," *Business Week,* May 29, 1989.
10. *New York Times,* August 8, 1992.
11. J. P. Womack, D. T. Jones, D. Roos, *The Machine that Changed the World,* Harper-Collins, New York, 1991.
12. P. F. Drucker, "The Emerging Theory of Manufacturing," *Harvard Business Review,* May-June 1990.

Index

Accountability, 27, 28, 30
Accounts payable, 3, 31, 34, 193, 194,
  195, 229, 241
  case study of, 196–201
Accounts receivable, 194
  case study of, 201–207
Activities, 18, 54–55, 89
Adaptability, 23, 112, 119
ALCOA, 121
American Society of Mechanical
  Engineers, 46, 83
American Society for Quality Control,
  ix, 9
Andreasen, P., 88
Apple Computer, 255
Appraisal costs, 69–70
Assembly drawings, 22–23
Assessment, 111, 112
Attribute lists, 35, 39
AT&T, 121, 142, 254
  four-stage model, xi, 160–163
Audits, 64–65, 111
Available time (complexity model), 96

Babbage, Charles, 13
Backward integration, 6
Baker, Edward M., 42
Baldrige Award (see Malcolm Baldrige
  National Quality Award)
Batch processes, 16, 233
Benchmarking, xi, 117, 121–122, 234
Best practices, 121, 122
Boedecker, R. F., 141
Boundaries, 16, 21, 31, 86, 129
Bureaucracy, 3, 8

Capacity, 23, 233
Chapple, Eliot, 236
Check sheets, 66, 135
Clumps, 250–251
Communication, 148
Companywide quality control (CWQC),
  10
Compartmentalization, 7
Competitive advantage, 8
Complaint and comment forms, 63
Complexity, defined, 95
Complexity model, 95–98, 110

Concurrent engineering, 209, 254
Conditions of excellence, 140, 141
Confidence intervals, 64
Conformance measurement, 66–67
Continuous improvement, 73, 145
Control, 240–241
Control chart, 22, 73, 75
Control points, 22, 59–60, 61, 132–133,
  241
Controlled processes, 60–63
Corrective action, 22, 77–79, 136–138
Cost measurement, 69–70, 242
Cost of quality, 69, 88, 114
Critical success factor concept, 72
Crosby, Philip, 10, 69, 147, 247
Cross-functional process, defined,
  18–19
Cross-functional process analysis, x,
  98–110
  process analysis technique, 99–105
  service blueprint, xi, 105–110
Current account, 117
Customer, defined, 33
Customer contact, 24
Customer-producer-supplier (CPS)
  model, 33–40
  applications of, 132, 218
  input requirement phase, 40
  output recruitment phase, 34
  production capability phase, 39–40
  in total quality management (TQM),
  145
Customer requirements, 131–132
Customer service check sheet, 65, 66
Customer surveys, 63–65, 66
  Bias, 63
  Direct customer surveys, 63
Cycle time, 68, 72, 78
Cycle time analysis, 213–215

DEC, 254
Decentralization, 25, 28, 142
Delivering Quality Service, 9
Deming, W. Edwards, 10, 142, 247, 251
Deming prize, ix
Department, defined, 87
Department activity analysis (DAA),
  83–93, 110

Departmental work product analysis, x, 87–98, 110
  complexity model, 95–98, 110
  department activity analysis (DAA), 88–93, 110
  department quality analysis (DQA), 93–95
Department quality analysis (DQA), 93–95
Deployment, 148
Deployment matrices, 35–39, 113, 145
Development, defined, 209
Development process:
  cycle time analysis, 213–215
  phases of, 210–213
Disaggregation, 151
Division of labor, 5
Documentation, 21–22, 34–39
Document distribution, xi, 18–19, 167
  example of, 128–138
Document of understanding (DOU), 100
Dottino, A. F., 88
Double blind surveys, 63
Downsizing, 253
Drucker, Peter, xi, 255–256

Education, quality in, 248
Effectiveness, 23, 112, 113–115
Efficiency, 3, 23, 112, 113–115
Employee involvement, 143–144, 244, 252, 253
Empowered work teams, 21, 28, 143, 144, 252, 253
Empowerment, 144, 244
Evaluation (see Process assessment)
Evaluation criteria, 111, 244
Excellence, conditions of, 140–141
Execution time, 109
External focus, 254–255

Fail points, 108–109
Failure costs, 69
Federal Express, 67, 145, 184
Federal Quality Institute, 248
Feedback, 22, 60, 77–79, 136–138
Feedback control, 16
Feigenbaum, Armand, 10, 69, 247
Finance, 193
Financial operations, defined, 193
Flat organization, 237, 240, 255
Flowcharts, 21, 35, 36, 46–48, 49, 50, 54
  drawing, 55–57

Flow diagrams, 45, 84–85
  (See also Flowcharts)
Flow-process charts, 45
Ford Motor Company, 142, 255
Fordham University, 142
Forward integration, 6
Fuller, F. T., 95

General Electric (GE), 68, 254
General Motors, 142, 253, 254
General Telephone & Electronics Corporation, 94

Harshbarger, R., 98
Hawthorne study, 230
Herbst, P. G., 243
Hewlett-Packard, 96
Hermann, Jaime, 42
Hidden plant, 7
Hierarchy model, 238, 239
Houghton, James R., 250
Human Factors, 242

IBM, 11, 73, 88, 98, 103, 116, 119, 121, 141, 142, 143, 145, 146, 216, 221, 253, 254, 255
Imai, Masaki, 139
Immediacy, 24
Improvement plan, 151
  Breakthrough, 214, 215
  Incremental, 214
Inbound logistics, 151
Industrial engineering, 13
Industrial Revolution, 5, 13
Inefficiency, 4–5, 7
  symptoms of, 114
Information flow charts, 35, 36
Informational transformation, 15
Input boundary, 31
Input-process-output model (see Transformation model)
Institute of Industrial Engineers, 46
Intangibility, 24, 61, 89, 147
Integrity, 193
Intel Corporation, 3
Interfaces, 18, 31–33, 86, 129–130, 237
  managing, 40–44
Intermittent processes, 16
Internal focus, 251–254
International Quality Study, 251
Ishikawa, K., 33

Job enlargement, 144
Johnson and Johnson, 254

Juran, Joseph, 10, 69, 145, 161, 247
Just-in-time (JIT) approach, 15

Kaizen, 145
Kaizen (Imai), 139
Kane, Edward S., 30
Kano model of customer satisfaction, 144
Kanter, Rosabeth Moss, 7
Killman, R. H., 231

Lean production, 254
Line of visibility, 108
L.L. Bean, 121
Locational transformation, 15, 184

Macro level diagram, 52
Malcolm Baldrige National Quality Award, ix, 9, 67, 121, 148, 248–249, 251–252
Management reporting process, 177–184
Mancuso, F. A., 222
Market share, 9, 72
Matrices:
    Deployment, 35–39
    Requirements-responsibility, 40–44, 99, 219–220
Mayo, Elton, 230
Mazda, 234, 255
McCabe, William J., 29, 73, 75
McKinsky and Company, 151
Measurement(s):
    of conformance, 66–67
    and corrective action, 77–79
    of cost, 69–70
    defined, 65–66
    example of, 71, 133–136
    graphical methods, 22, 72–77
    purpose of, x, 22
    of repetition, 69
    of response time, 67–68
    selection and implementation considerations, 70–72
    of service level, 68
    in a service organization, 22
Memorandum of understanding (MOU), 100
Metrics, 121
Milliken, 121
MIT Commission on Industrial Productivity, 7
Mitsubishi, 255

Motorola, 67, 146
Multifactor productivity, 113

National Cash Register (NCR) Corporation:
    five-dimension model of process management, xi, 156–160
Non-accumulation, 24
Nonconformance, 67, 69
Nordstrom, Jan, 89

On the Economy of Machinery and Manufactures, 13
Operation sheets, 45
Ordering process, 167–177
Organization, evolution of, 5–8
Organizational barriers, 7
Organizational structure, work flow and, 24–26, 236–237
Organizational trends, 249–256
    external focus, 254–255
    internal focus, 251–254
    new view of, 255–256
Outbound logistics, 151
Output, 14
Output boundary, 31
Outsourcing, 254
Ownership, 21, 22, 27–31, 128–129

Parallel activities, 238, 240
PAT, 98, 99–105
Participative management, 147
Performance evaluation, xi, 120
Physical transformation, 15
Prevention costs, 70
Primary activities, 151
Problem-solving work teams, 143
Procedures, 45
Process:
    characteristics of, 20–23
    defined, 11, 14, 15, 18
    elements of, 14–15
    manufacturing and service compared, 22, 23
    origins of, 13–14
Process analysis, 81–82
    application example, 82–86
    basic questions of, 83
    cross-functional process analysis, 98–110
    defined, 81–82

Process analysis (continued)
  departmental work product analysis,
    87–98
  procedure, 82
  service blueprint, xi, 105–110
  traditional methods of, x, 82
Process analysis technique (PAT), 98,
    99–105
  steps in, 100–105
Process assessment:
  benchmarking, xi, 117, 121–122, 234
  criteria for, 112–115
  performance evaluation, 120
  process capability, 120
  quality profile, xi, 122–124
  questionnaire, 125–126
  rating method, 115–119
  reasons for, 111–112
Process block, 52
Process capability, 120
Process charts, 45
Process control, x, 59–79
  control points, 59–60
  feedback and corrective action, 77–79
  measurements, 65–70
  reactive and controlled processes,
    60–63
  in the service process, 63–65
Process costs, 242
Process data sheet, 52
Process definition, x, 45–57, 78
  application example of, 130–131
  defined, 45
  drawing flow charts, 55–57
  flow charts and symbols in, 46–48
  step-by-step method of, 48–55
Process design, x, xi, 229–243
  activity, categories of, 238–242
  boundaries and ownership
    information requirements, 235
  constructing a process, 237–244
  hierarchy model, 238
  human factors, 242–244
  process costs, 242
  requirements for, 230–234
  work flow, 234–237
Process evaluation, rating method,
    115–119
Process improvement:
  ownership and, 30–31
  prevalence of, 251
  in process management, 154–155
Process improvement plan, 151, 153

Process initialization, x, 27–44, 78
  boundaries and interfaces, 31–33
  customer-producer-supplier (CPS)
    model, 33–40
  managing the interface, 40–44
  process ownership, 27–31
Process management:
  basic steps in, 127
  boundaries and interfaces, 129–130
  control points, 132–133
  customer requirements, 131–132
  defined, 11,
  defining the process, 130–131
  evolution of, ix, 20
  feedback and corrective action,
    136–138
  in financial operations, 193–207
  five-dimension model of at NCR, xi,
    156–160
  four-stage model of at AT&T, xi,
    160–163
  implementation, examples of,
    155–163
  improvement through, 149–163
  in a laboratory, 209–225
  measurement and assessment,
    133–136
  ownership, 128–129
  repeatability, 16
  in staff/service operations, 167–192
Process management analysis, 154
  example of, 184–192
Process ownership (see Ownership)
Process quality management and
    improvement (PQMI), 160, 161
Process process charts (PPCs), 45
Process reengineering, 155, 233–234
Process selection, 150–151
Process teams, 153–154
  approach to improvement, case study
    of, 221–225
Procter and Gamble, 254
Production capability, 39–40
Productivity, 3, 7
  measures of, 113
  multifactor, 113
Product Development, 151
Product-process matrix, 232–233
Profit analysis, 109
Project management, 25–26, 42

Qualification, 215–221
Quality awareness training, 147

Quality circles, 143
Quality costs, 69
Quality function deployment (QFD), 39,
147, 210–211
Quality improvement, 247
and market share, 9
service quality, 215–221
through process management,
149–163
and total quality management
(TQM), 10–12, 145–146
Quality improvement teams (QIT), 143,
156, 252
Quality inspection, 60–61
Quality management team (QMT), 156
Quality profile, 122–124
Quality revolution, 247–249

Rating method, 115–119
example of, 117–119
Reactive processes, 60–63
Reciprocal activities, 238, 240
Recognition, 147–148
Reengineering, 155, 233–234
Regulation, x, 22
Repeatability, 16
Repetition, 69
Requirements-responsibility matrix,
40–44, 99, 219–220
Response time measurement, 67–68
Reward system, 5, 25
Rockhart, N., 72
Roethlisberger, F. S., 230, 235
Rosecrance, Richard, 249
Routings, 22, 45

Samford University, 142
Saturn project, 252
Sayles, Leonard, 29, 236
Scientific management, 13, 46, 81, 230
Segmentalism, 7
Segmentation, 7, 8
Self-directed work teams (see
Empowered work teams)
Sequential activities, 238, 240
Service blueprint, xi, 105–110
Service industry, evolution of, 8
Service level agreement, 68
Service level measurement, 68
Service process(es), 15, 23–24
features of, 23
improving service quality, methods
of, 215–221

Service process control, methods of,
63–65
Shostack, G. Lynn, 105
Simultaneous engineering, 209, 254
Six Sigma Program, 146
Smith, Adam, 5
Smith's Automotive Products Company,
252–253
Sociotechnical analysis, 230
Sociotechnical theory, 230
Span of control, 237
Specialization, 5, 7, 236
Specifications, 35, 39
Statistical process control, 63
Statistical quality control, 247
Strategy, 148–149
Suboptimization, x, 7, 8, 10, 25, 209,
237, 250
Subprocess, 18, 19–20, 33
Support activities, 151
Support services, 8, 254
Surveys, 63
Symbols, 46–48, 49, 50, 51, 83
System disaggregation, 150–151
Systems process view, 256

Target ratcheting, 73, 145
Task(s), 18, 53, 89–90, 241
Task analysis, 86
Task-dependency matrix, 243
Task-oriented work teams, 143–144
Tavistock Institute, 230
Taylor, Frederick W., 13, 230
Telephone surveys, 65, 66
Thompson, S. D., 236, 240
Tiger teams, 143, 254
Time and motion studies, 13
Timeliness, 194, 241
Toshiba, 255
Total Quality Control (Feigenbaum), 10
Total quality control (TQC), 10–11
Total quality management (TQM):
assessing progress, 146
communication, 148
continuous improvement, 145–146
customer focus, 144–145
employee involvement in, 143–144
and excellence, 140–141
features of, 139–140
growth of, 3, 247
improvement through process
management, 149–163
management's role in, 141–142

Total quality management (continued)
  organization-wide involvement,
    142–143
  process approach to, 150–155
  recognition, 147–148
  strategy and deployment, 148–149
  training, 147
Training, 147, 244
Transactional transformation, 15
Transformation, 14–15, 16, 240–241
Transformation model, 14, 16–20
Transportation service process, analysis
  of, 184–192

Unavailable time, 95–96
United Parcel Service (UPS), 184
U.S. Postal Service, 253
Unit operations, 46
University of Pennsylvania, 142
University of Rhode Island, 142

Value chain method, 151, 152
Verification, x, 22
Volvo, 144

Waste, 3–5, 7
Wealth of Nations, The, 5
Westinghouse, 70, 140
Windsor Export Service (WES), 196,
  229
Word descriptions, 35, 45
Work components, examples of, 96
Work flow, 7, 21, 24–26, 234–237
Work group, 87

Xerox Corporation, 9, 121

Zero-defect program, 10, 145, 146

ABOUT THE AUTHOR

Eugene H. Melan is a visiting associate professor of Business at Marist College where he specializes in teaching production and operations management after an extensive and varied career at IBM. While at IBM he was engaged in a number of firsts with the company in technical and management positions—the first generation computer, the first generation computer memory and integrated circuit technologies, and the first quality improvement efforts. As a program manager for quality improvement, he was deeply involved in developing, teaching, and implementing quality improvement methods since its inception in the company. He also serves as a consultant to major organizations such as Bell Laboratories, Dupont, MCI, and the University of Pennsylvania. The author of numerous professional papers and articles, Mr. Melan is a fellow of the American Society for Quality Control.